Mental Illness

Mental Illness

THE JOURNEY'S END

Leonard C. Wilton

Copyright © 2017 by Leonard C. Wilton.

ISBN: Softcover   978-1-5434-2210-8
             eBook   978-1-5434-2209-2

All rights reserved. No part of this book may be reproduced or transmitted in any form or by any means, electronic or mechanical, including photocopying, recording, or by any information storage and retrieval system, without permission in writing from the copyright owner.

Any people depicted in stock imagery provided by Thinkstock are models, and such images are being used for illustrative purposes only.
Certain stock imagery © Thinkstock.

Print information available on the last page.

Rev. date: 05/12/2017

To order additional copies of this book, contact:
Xlibris
1-888-795-4274
www.Xlibris.com
Orders@Xlibris.com
760739

Dedicated to mom and dad.
They were always a phone call away.

Foreword

"Mental Illness: The Journey's End" is interesting, informal, quasi-technical, and an inspiration to readers. The format and style are simple and straightforward, and the terminology (commentaries and definitions) are clear and accurate.

The subject matter is treated carefully and thoroughly, and where appropriate are commentaries that explain terms, cite examples, and allow us to empathize with the disorder, understand the scientific basis and learn more about the topic.

By reading this book in its entirety or by scanning up and down one can gain insight, learn about, and understand the many different forms and treatments.

Stephen Clements

It is the journey that illumines us so our destinies we cherish.

Stephen Clements

Mental Illness: The Journey's End

Descartes talked of six passions, James four, and Watson the basic three (Garrison, 1948: 53). Presented by Watson, these emotions were sufficiently differentiated in respect to a volume of research on the subject of the three original emotions -- love, fear, and anger (1948 : 53).

In discussing love, "the emotion of love is directly related to the sexual impulse. . . . [and] is the consequence of physiological disturbances" (1948 : 53). In 1933 A.R. Gilliland stated that : "The earliest loves are not sexual in character, Freud to the contrary notwithstanding. However, sex stimulating gives pleasure and becomes a large factor in love responses" (qtd. in Garrison, 1948 : 53).

Although many great minds were ministerial in advancing the field of psychiatry, Freud was the most recognized and influential; his praxis, methodologies, and analyses provided a prototype for a myriad of professionals to follow and emulate. Freud interpreted and treated everything from childhood trauma and maladjustment, to neuroses, psychoses, hysteria, disturbing dreams, and repressed emotionalism.

His treatise on human sexuality was daring, innovative, and tenable. Freud paved the way for those stricken with mental illness to be administered to psychiatrically, his theories and analyses bringing to light and treating the afflicted's conscious and subconscious dissonance. By probing into the cimmerian, subterranean recesses of the unconscious psyche, a professional could remedy the neurotic symptomatologies, thereby offering the patient a more salubrious psychic conception.

Mental illness is accountable for the most admissions in hospitals, with the highest readmissions.

Hospitals: The Private Hospital

At issue here is the private hospital with the psychiatrist charging $90 an hour to listen to a patient discuss many such issues from a meaningless dream to family problems. These patients are safeguarded from lower level care, and although they are initially locked on a ward their surroundings are to a significant degree more lucullan. There are more therapies, group therapies, and discourse between patient and professionals.

Private institutions are frequented by the wealthy as well as the destitute (those with public insurance). Some spend from a week to a year hospitalized; others leave on a daily basis to acquire training or education for employment. Others return to their families or are placed in convalescent or board and care facilities.

The psychological care in these private hospitals is excellent. Many aides, nurses, social workers, and rehabilitation workers have years of experience and have a good rapport with the patients. There is also a lower turnover rate and more time can be set aside to individual care.

Some group therapies in these settings tend to engage superficial and insignificant matters, however. For example, one session may involve such trivialities as preparing frozen food or visiting a museum.

Illnesses range from mild depression to anxiety and adjustment deficits, nervousness and compulsive disorders to delinquency in younger patients. Group therapies tend to identify problems, connect patients interactively, and relieve some burdensome and conflictual bugbears.

I have encountered some unusual cases. There was an older woman admitted to Cedars-Sinai Thalians Psychiatric Unit where I spent four months. She was depressed. A married woman with a son studying at Berkeley, she was overly bright. She was extremely withdrawn but during a large group session the topic of sex came to light. There was a resounding reaction. This woman, rising to the occasion, became involved in the discussion, and it inarguably had a therapeutic and remedial effect. It had actually generated some enthusiasm, probably a maternal and amatory proclivity with which she'd had a positive experience. This discussion provided the woman with hope and the feeling of renewal.

Psychic Growth

Throughout the world of psychiatry and psychology, there are different theories explaining childhood and adolescent development. Many agree that values are acquired through pubescence and are formed into an individual's sense of social decorum. Professionals also maintain the superego is acquired genetically. Despite these theories, adolescence is a psychological as well as an emotional metamorphosis and many youths find gratifying expression as a necessary cog in the machinery of socio-cum-emotional development.

Many kids can be cruel especially among those with a rebellious mien. Everything is flawed and only the self is note perfect. These are the days of ego defense and development. The child is neither emotionally healthy nor troubled. The adolescent in conflict is likely to be spiteful, unappreciative, and unmanageable.

There are numerous examples where teens become involved in gangs, crime, and are headed for legal troubles later in life. The emotionally healthy student is involved in extra-curricular activities and is socially adequate. These individuals are goal oriented and look forward to a bright future.

Social aptitude is important although proper nurturing is necessitous. The family unit breeds family units. Parents need to be authoritative yet promote what is deemed caring and supportive. A child undisciplined will find himself criticizing others and be likely to acquire poor character.

Stages of Early Development

The child is born requiring much love and care. Its initial stimulation is oral. A baby develops pleasure from this stage and the primary drive is in acquiring nourishment, the feeding being manifested through these means. Soon the child is exploring outside its eccentric realm and is able to perceive, learn, and grow. Many believe that during these formative years the foundation is constructed in their life-long edifice. A stable environment begets a stable individual. Impressions are especially important here. One's emotional and psychological well-being hangs in the balance.

Children at these stages undergo various learning progressions starting with an oral one. Next, an anal stage is experienced by the toddler. A pleasurable component of the stimulation is in the act and actual smearing of the feces, the child learning to acquire possessions and enjoy the satisfaction of having them later in development. It is here that the child learns to give objects to those in proximity. There is an element of altruism and generosity such that later in life one who has had a positive anal stage of development will become philanthropic and bestow gifts to others (Jastrow, 1946 : 217).

There are three characteristics associated with the anal-erotic course of development. The first is orderliness which may lead to pedantry. The second, parsimony, may develop into avarice. The third characteristic in this triad is obstinacy which may evolve into vindictiveness and irascibility (1946 : 216).

Enjoyment is a likelihood here, as well as there being a sense of relief in performing an excremental activity. "Anal" people, according to Freud, tend to be conservative while "oral" are liberal, politically speaking (1946 : 217).

Environmental Factors: How Valid?

Another school of thought and reason professes that environment is not a factor in defining a person's mental health later in life. Consider the classic cases of Anne Frank and Charles "Tex" Watson. Although Anne Frank endured a traumatizing and dreadful childhood, she still expressed hope and love for humanity. And in the case of Charles

Watson, this high-achieving, well-adjusted student joined a cult and committed murder. How are these two antipodal exempla justified?

A Social Phenomenon

Psychology also teaches us that every scenario is different, individual in effect, that people are fashioned diversely. Sociological expedients are a consideration as well as individual psycho-social development. Everything boils down to a socially-related circumstance. It is others we want to impress. The individual acquires social, job-related, and monetary status to gain the approval of others. Security, both financial and emotional, are the ends to these means. Society measures success according to these standards -- unfortunately without considering spiritual or other intangible factors which may lend themselves a more valuable asset in their pursuit of happiness and well-being.

"A man is no loser who has friends" is a classic line from a classic film. The man or woman who has character, honor, and a benevolent disposition is wealthy in the most true sense of the word.

Science: A Testimony

While Freud promulgated the rudiments of sexuality, Carl Jung espoused the basic drive in acquiring food as primordial. Many philosophies, religions, and rituals had their particular reasoning throughout the ages. Even astrology tried to explain phenomena whereas other means were not useful in consideration.

Basic psychological functioning can be treated as a myriad of interrelated processes. Some may rely on feeling or intuition, alone or coupled with sensation. Some people are extremely sensitive while others are aloof and unaffected. This is partly why individuals, I feel, should be identified in different ways; the old generalizing and stereotyping are not instrumental or practical.

We know that the mind is a psychological entity. What are the incorporeal manifestations of that bodily organ? Rhetorically and scientifically speaking, that process has perplexed humanity from antiquity. Many believed the mind -- the ability to think, reason, remember, and communicate -- was what separated man from animals. Others believed in a soul, a covert intangibility. But these beliefs did not explain the mental processes. How does substance incorporate and embody thought or memory? Is it a sense? Can it know the future? Is it a divine attribute?

The Early Institution

Mental hospitals have been around since our awareness of mental illness. Insane asylums were constructed to house and protect patients, many of whom were confined to them indefinitely. The treatment, care, and quality of living conditions in these asylums were horrendous. It is well-known that abuse and mistreatment went unchecked. There were no agencies responsible for maintaining fair and decent quality of care, and there was no way to wire Washington for help.

At the onset of this century, great strides were undertaken to understand, treat, and administer humane and proper care. Great minds like James, Freud, Breuer, Adler, Sullivan, Maslow, Fromm, Jung, and Menninger served by scientifically identifying and consequently offering techniques to remedy this terrifying illness.

Before modern treatments were available, some families tried to deal with an afflicted relative by secluding them in a cellar or attic, the prevailing notion that a demonic possession was inducing this bizarre behavior. There are cases of these so-called possessed and their families that alluded to them, for example, as the uncle who liked to be alone or the grandmother who liked to stand in the dark and vent feelings. The understanding of these individuals' families was as irrational as the afflicted person's behavior.

Today institutions are managed by agencies that insure that treatment is humane and appropriate. Although psychiatric care is not an infallible praxis, there are medical standards to maintain and mental hospitals are also amenable to legal jurisprudence.

Pain v. Pleasure

I read in adolescent psychology that pain is relief from an uncomfortable situation and this pain can be considered as pleasant (Garrison, 1948 : 48). "That observation is an synchrony with, for example, the conviction that growth is an unpleasant reaction to a stimulus. But one is likely to feel pain as a reaction to many unpleasant stimuli. Here it should be noted that feelings are often defined as pleasure or pain. Pleasure is associated with man's affective life and pain refers to sensations" (1948 : 48).

Feeling, however, is independent of stimulation and is correlated with a fundamental attitude existing in the organism. These must be recognized and studied. It is a mistake according to Henry T. Finck for social reformers to follow the Puritans and taboo all pleasures. For an organism to flourish, pleasure is intrinsically necessary (1948 : 48).

Onanism

Many people gratify themselves while fantasizing about different erogenous zones and/or erotic experiences. The term autoerotic applies to a sexual proclivity that is tendentiously egocentric. Instead of engaging in a group or partnered activity, the autoerotic, whether a male or a female, wants an immediate and enjoyable fantasy fulfilled.

It is interesting how many individuals fantasize about sex. Some fancy being walked on by a group of women or tied and bound and humiliated. Domination in sexual matters can include anything from fascination to vivisection.

The Encephalon

The mind is composed of countless nerve cells and nerve fibers. The three basic divisions are: the cerebrum, the cerebellum, and the medulla oblongata.

The cerebrum with its two hemispheres is thought to control conscious and voluntary processes such as thinking and body movements (Webster's, 1997 : 229). The cerebellum, with its two lateral lobes and middle lobe, function as a center for muscular movements (1997 : 229). The medulla oblongata, as the continuation of the spinal cord, contains nerve centers that control breathing, circulation, etc. (1997 : 843).

There are many other processes:

"Diencephalon: the posterior end of the forebrain, including the thalami and hypothalami" (1997 : 343).

"Thalamus: a mass of gray matter forming the lateral walls of the diencephalon (thalamencephalon) and involved in the transmission and integration of certain sensations" (1997 : 1385).

"Hypothalamus: the part of the diencephalon in the brain that forms the floor of the third ventricle and regulates many basic bodily functions, [such] as temperature" (1997 : 665).

"Hippocampus: a ridge along the lower section of each lateral ventricle of the brain" (1997 : 639).

"Pons Varolii: a piece of connecting tissue; spec., the bridge of white matter at the base of the brain, containing neural connections between the cerebrum, cerebellum, and medulla oblongata" (1997 : 1049).

"Pia Mater: the vascular membrane immediately enveloping the brain and spinal cord and surrounded by the arachnoid and dura mater" (1997 : 1020).

"Dura Mater: the outermost, toughest, and most fibrous of the three membranes covering the brain and spinal cord" (1997 : 422).

"Arachnoid: designating the middle of the three membranes covering the brain and spinal cord" (1997 : 69).

"Pituitary Gland: a small, oval endocrine gland attached by the stalk to the base of the brain and consisting of an anterior and of a posterior lobe; it secretes hormones influencing body growth, metabolism, and the activity of other endocrine glands, etc." (1997 : 1030).

"Pineal Body: a small, grayish, cone-shaped, glandular outgrowth from the brain of all vertebrates that produces the hormone melatonin; in lower vertebrates, it is often visible as an external median eye" (1997 : 1026).

"Melatonin: a hormone, $Cl2\ Hl6\ N2\ O2$, produced by the pineal body, that lightens up skin pigmentation, inhibits estrus, etc.; its secretion inhibited by sunlight" (1997 : 844)

"Corpus Callosum: a mass of white transverse fibers connecting the cerebral hemispheres in humans and other higher animals" (1997 : 312).

"Corpus Striatum: either of two striated ganglia in front of the thalamus in each half of the brain" (1997 : 312).

Other functions, processes, and systems include: neurotransmitters, the limbic system, neuromodulators, sulci, ventricles, dendritic activity, sylvan fissure, meninges, and frontal, temporal, parietal, and occipital lobes.

Brain damage, synaptic misfiring, and deadened nerve cell and fiber are considered irreversible, although some functioning can improve, the mind healing itself. I have my particular theory as to how this occurs: dendritic embranchment. I feel the nerve cell can ramify and restore some inert tissue, to some extent, to activate cellular functioning.

Psychology: Its Goals and Perspectives

Between the 1920s and 1950s psychologists defined the field of psychology -- the discipline -- as a study of behavior. But when under attack it evolved into a science of thinking, dreaming, and everything else that went on between people's ears (Wade & Tavris, 1993 : 4). An appropriate definition of psychology is "the scientific study of behavior and mental processes and how they are affected by an organism's physical state, mental state, and external environment" (1993 : 4). Simply put, its goals are to "(1) describe, (2) understand, (3) predict, (4) control or modify behavior" (1993 : 4).

Some sciences that are closely related but are different fields of inquiry are:

"Sociology: the study of groups and institutions in society and includes the family, religious institutions, the workplace, and social cliques. Social psychology falls on the border between psychology and sociology" (1993 : 7), and studies how social groups and situations affect behavior and vice versa.

"Anthropology: the physical and cultural origins and development of the human species" (1993 : 7). In the prosperous field of cross-cultural psychology, professionals are studying differences and similarities among cultures (1993 : 7).

Economics and political science are two other related disciplines albeit they are only incidentally tied to the behavioral aspect of psychology (1993 : 7).

The five perspectives or schools of psychology and their definitions are:

> "(1) Behavioral: an approach emphasizing objectively observable behavior and how the environment influences, plays a role in human or animal behavior" (1993 : 13).
>
> "2) Psycho-dynamic: refers to the unconscious energy dynamics in an individual, such as inner forces, conflicts, and instinctual energy" (1997 : 15).
>
> "(3) Cognitive: emphasizes the importance of mental processes in areas of perception, memory, language, problem solving, and other behaviors" (1993 : 16).
>
> "(4) Physiological : treats bodily events and changes associated with actions, feelings, and thoughts" (1993 : 17).
>
> "(5) Sociocultural perspective: studies how social and cultural influences affect behavior" (1993 : 19).

Contemporary Paradigms in Psychopathology

Varying theories and their attendant etiologies and some treatments are as follows:

The Behavioral Paradigm theorizes that abnormal behavior is a manifestation of aberrant somatic or physiological functioning (Davison & Neale, 1996 : 28). Some research suggests, for example, that schizophrenia is attributable to heritable causes, depression is caused by unusual neural transmissions, and anxiety disorders stem from a defective autonomous nervous system (1996 : 29). In consideration of organic mental diseases, impairment in brain structure is etiological. These disorders are thus treatable by proper administration of medications.

The Psychoanalytic Paradigm was introduced by Sigmund Freud (1836 - 1939), and ascribes that psychopathology results from unconscious conflict (Davison & Neale, 1996 : 33). Some of his theories on the structure of the mind include: the id, which reflects basic urges for survival and pleasure; the ego, which mediates between demands of reality and gratification, and the superego, which is acquired from the child's parents and serves as the vehicle of society's moral standards (1996: 34). The human personality as a closed energy system has a determinate amount of energy, with each part of the personality also battling for a share of this energy. Every bit of behavior has specific causes, many of these unconscious (1996 : 35).

The Learning Paradigm professes that abnormal behavior, like other human behavior, is learned, or acquired environmentally. Initially experimental psychologists Wilhelm Wunt (1882-1920) and Edward Titchener (1867-1927) implemented procedures to study the effects of stimuli on subjects who reported on their experience (Davison & Neale, 1996 : 43). "Through painstaking introspections subjects attempted to uncover the building blocks of experience and the structure of consciousness" (1996 : 43).

The introspectionism, however, lost credibility as different laboratories were yielding conflicting data. The resultant behaviorism that ensued, defined as the study of observable behavior rather than consciousness, gained tangible acceptance.

John B. Watson (1878-1958) revolutionized psychology by rejecting introspectionism -- and through his efforts changed the focus of the science from thinking to learning. Ivan Pavlov (1849-1936), psychologist and Nobel Laureate, achieved high honors with his extensive research and theories in classical conditioning (Davison & Neale, 1996). And a renowned behaviorist, B.F. Skinner, in one of his best utopian novels, Walden Two, "argued that freedom of choice is a myth and that all behavior is determined by the reinforcers provided by the social environment" (1996 : 46). Skinner also did much to revolutionize the science, and his theories are the harbingers of contemporary behaviorism.

"The Cognitive Paradigm groups together the mental activities of perceiving, recognizing, conceiving, judging, and reasoning, and focuses on an individual's experiences, how they are useful, and in transforming environmental stimuli into information that is also useful" (1956 : 50). According to cognitive psychologists, the learning process is more complex than forming new stimulus-response associations. They view even classical conditioning "as an active process by which organisms learn about relationships among events rather than an automatic stamping-in of associations between stimuli" (Rescorla, 1988, cited in Davison & Neale, 1996 : 50). "The learner fits new information into an organized network of already-accumulated knowledge, often referred to as a schema according to Neisser in his 1976 study (cited in Davison & Neale, 1996 : 50). "Schema is a mental structure for organizing information about the world" (Davison & Neale, 1996, glossary : 22). New information may fit the schema or, if not, the learner reorganizes the schema to fit the information. Scientific paradigms resemble a cognitive schema in that they screen our experiences (1996 : 50).

Prominent cognitive psychologists Albert Ellis and Aaron Beck have made positive contributions in the field. Ellis "has focused on the role of irrational beliefs as causes of abnormal behavior" and Beck "developed a cognitive therapy for the cognitive biases of depressed people" (1996 : 52).

"[The] consequences of adopting a paradigm include a decision concerning what kind of data will be collected and how they will be interpreted. This approach could overlook other information in advancing what seems to be the most probable explanation" (1996 : 53).

A behaviorist would attribute the high prevalence of schizophrenia in lower-class groups to a paucity of social rewards or positive reinforcement. A biologically-oriented theorist would challenge this consideration saying that many deprived individuals do not become schizophrenics (1996 : 53).

Another paradigm that merits consideration is called the Diathesis-Stress Paradigm. It includes biological, psychological, and environmental factors. Diatheses refers to a constitutional predisposition toward illness and stress refers to environmental or life disturbances (1996 : 53-54). For example, on a psychological level, the chronic feeling of hopelessness may be considered a diathesis for depression.

All these paradigms present treatment approaches which a behaviorist may implement -- prescribe medications, for example -- or a psychoanalyst may consider opportunities for behavior modification. "Most therapists, however, subscribe to eclecticism and employ ideas and protocols from a variety of schools as determined by Garfield & Kurtz in 1974" (Davison & Neale, 1996 : 55).

Unhealthy Behavior Patterns

Life is full of responsibilities -- paying rent, helping the kids with their homework, getting to work on time, investing in the right fund. Little wonder we are worried, apprehensive, stressed, depressed, cloistered, and neurotic. Every positive achievement is matched with something negative. No matter how hard we try we cannot have everyone approve of us. There is little room to sustain oneself on the precarious path of life. If we veer to the left, we are weak and incapable. If we swerve to the right, we are thought of as malicious and selfish.

There is a such a volume of unhealthy reactions to every situation, normal behaviors become injurious. We feel headaches, develop ulcers, have nervous breakdowns, contemplate suicide. Our emotions we are unable to compromise. Life becomes frightful, out of control. We want escape. We swallow a bottle of sleeping pills.

Now this is a sample of a scenario that is possibly the worst nightmare into which living can evolve. Although there are many problems, there are many solutions. We can opt out of this nightmare and lead a healthy life, be happy, and avoid the troubles that render us neurotics and psychiatric cases.

Anger

Expressing anger is indeed easier than trying to suppress it, albeit even healthier is not having the anger at all (Dyer, 1976 : 210). Some alternatives to feeling and containing this emotion are:

(1) Realizing that you can change your thoughts.
(2) Postponing the anger for a few minutes as a mode of controlling it.
(3) Reminding yourself that anyone can choose to be what they choose and not making demands on others' behavior.
(4) Communicating with those that provoke this anger to arrive at a mutual understanding in order to check and ward off any future episodes or incidents.
(5) Understanding that everything you believe will be met with disapproval by 50 percent of everyone, and expect that (1976 : 219-20).

"Instead of being an emotional slave to every frustrating circumstance, use the situation as a challenge to change it" (1976 : 221).

Breaking Free from the Past

We as human beings evaluate ourselves in terms of self-descriptors. These criteria are not inappropriate, but can cause harm (1976 : 75). All self-labels come from an individual's history, and the past -- as Carl Sandburg said in Prairie -- "is a bucket of ashes" (qtd. in Dyer, 1976 : 75).

Tags such as "I am shy"; "I am nervous"; "I am lazy" serve to keep us in the past and inhibit growth. "When an individual must live up to the label, the self ceases to exist" (Dyer, 1976 : 75). This is what Soren Kirkegaard meant when he wrote, "Once you label me, you negate me" (qtd. in Dyer, 1976 : 75).

In eliminating some of the vexing "I'ms" that keep us looking back and living in the past, we could try (1) substituting "I'm" with sentences that imply that this is the way "I" used to be, or this is the way "I" used to label myself. The four neurotic sentences:

(1) "That's me."
(2) "I've always been that way."
(3) "I can't help it.
(4) "That's my nature" (Dyer, 1976 : 75)

can be changed to

(1) "That was me."
(2) "I can change that if I work on it."
(3) "I'm going to be different."
(4) "That's what I used to believe was my nature." (Dyer, 1976 : 86).

Get involved in an activity, one you avoided in the past, and after your three-hour immersion ask yourself if you are the same "I'm" you used this morning (1976 : 86).

The Useless Emotions: Guilt and Worry

Throughout life, the two most otiose emotions are guilt for what has been done, and worry about what might be done. Guilt and worry force us to look backward or forward and as a result we squander the present moment (Dyer, 1976 : 89-90). Guilt is the most impracticable emotion, expending "by far the greatest waste of emotional energy" (1976 : 91).

In eliminating guilt, here are some strategies:

(1) "View the past as something that cannot be changed.
(2) Ask yourself why we are avoiding something in the present with guilt about the past" (Dyer, 1976 : 102-3).
(3) Accept things about yourself that others may not.
(4) Assess the consequences of your behavior. Are the consequences of your actions pleasing and productive for you? (1976: 103).

Worry

Just as guilt is fostered by society, so too it encourages worry. Worry is a byproduct of our culture and it will most likely render one less effective in dealing with the present. Most individuals worry about things they cannot control, and this foible is all the more onerous.

Justifications for worrying include a desire to:

(1) Escape for the present moment.
(2) Avoid risks by using worry for immobility.
(3) Label yourself as caring by worrying (1976 : 112-13).

Some strategies for obviating worry include a conscious commitment to:

(1) View the present as the time to live, not the future.
(2) Realize the futility of worrying. Can anything change as a result?
(3) Face the fears you have with productive thought and behavior (1976 : 113-15).

Jealousy: A "Demand for Justice" Sideshow

John Dryden called jealousy "the jaundice of the soul" (qtd. in Dyer, 1976 : 168). If jealousy gets in your way and causes any emotional immobility, then it should be eliminated. The definition of jealousy is actually a demand that someone love you in a certain way, and if that someone does not love you, you react by feeling rejected, estranged. It is a result of a lack of confidence as it is an other-oriented activity (Dyer, 1976 : 168).

Jealousy allows another's behavior to be the cause of your emotional discomfort. People who really like themselves do not choose jealousy or allow themselves to be distraught when someone else does not play fair (1976 : 168).

Procrastination

There are three neurotic phrases that enable us to rationalize putting things off:
"I hope things work out.
I wish things were better.
Maybe it will be okay" (Dyer, 1976 : 177).

Individuals can accomplish anything they set out to do, but by defending action they allow escapism, self-doubt, and self-delusion to prevail. Wishing and hoping are the stuff of fairy tales. By moving away from being strong in the now, you only look toward the future hoping things will improve (Dyer, 1976 : 177).

The Folly of Shoulds, Musts, and Oughts

Albert Ellis coined a word, "musterbation," that defined the tendency to incorporate "shoulds" in one's behavior. While you are behaving in ways you feel you must, rather than some other form of behavior, you are musterbating (Dyer, 1976 : 148).

The brilliant psychiatrist Karen Horney comments on this topic in "The Tyranny of Should," as follows: "The shoulds always produce a feeling of strain, which is all the greater the more a person tries to actualize his shoulds in his behavior. Furthermore, because of externalizations, shoulds always contribute to disturbances in human relations in one way or another" (qtd. in Dyer, 1976 : 148).

In addition to all the shoulds we arrogate, there are many should-nots. These include some of the more obvious: "you should not be rude, angry, foolish, silly, juvenile, lewd, gloomy, offensive, and many others" (Dyer, 1976 : 148). Any should or should not will have to produce a strain on you, since you will not be able to fulfill your erroneous expectation. The strain referred to does not result from the "must behavior" but from the imposition of the should (Dyer, 1976 : 148).

Approval Seeking

According to Dyer, "when approval seeking is a need, the possibilities for truth are all but wiped away. If you must be lauded, and you send out these kinds of signals, then no one can deal with you straight. Nor can you state with confidence what it is that you think and feel at any present moment of your life. Your self is sacrificed to the opinions and predilections of others (Dyer, 1976 : 51).

Seasonal Affective Disorder

This form of depression is differentiated from other depressions as it is almost exclusively due to external factors. SAD is characterized by the sudden onset of illness (depression) during seasonal changes of sunlight (The Life Extension Foundation, 1998 : 205). Apparently loss of sunlight induces chemical changes in the brain and is causal (1998 : 205).

Individuals affected differ in that those disposed have a characteristic brain, with evidence of serotonergic dysregulation involved in SAD (1998 : 205).

Bright light therapy is effective in treating SAD (1998 : 205). Prozac helps also. A safer alternative in boosting serotonin levels is ingesting doses of tryptophan.

Vertigo

More a condition than an illness, "vertigo involves feelings of dizziness, faintness, and the inability to maintain normal balance while sitting or standing" (1998 : 219).

Etiologically, ear infections, ear surgery, or injury can cause vertigo. There may be a neurological deficit extant that requires treatment if conventional medication is not efficacious (1998 : 219).

Hydergine can be effective in treating vertigo, as is piracetam. In a recent study gingko extract helped improve symptomatologies manifested by a cerebral circulatory deficit (1998 : 219).

Insomnia

There are a myriad of factors that are accountable for insomnia, vis-a...-vis those that affect persons entering middle age and beyond. Generally these individuals are affected due to the deficiency of the hormone melatonin (1998 : 142).

"Melatonin is released by the pineal gland, induces drowsiness, and allows the body to enter deep-sleep patterns" (Life Extension Foundation, 1998 : 142).

Treatment protocols include melatonin supplementation, and vitamin B-12 to normalize circadian rhythms. It should be noted here that valerian taken over an extended period of time has a significant toxicity risk factor (1998 : 142-3).

Cognitive Enhancement, Aging

Aging causes us to decline in overall cognitive functioning ability. Our capacity to store and retrieve short-term memories and to learn new information becomes impaired. Many neurological diseases develop from aging (Life Extension Foundation, 1998 : 5).

Studies demonstrate that brain aging and its debilitating effects can at least be moderately controlled. Acetylcholine precursors choline, lecithin, and phosphatidylcholine improve the effects of aging (1998 : 5). "Acetylcholine, a neurotransmitter, helps brain cells communicate with each other" (1998 : 5).

Dysphoria

The etymon from the "Greek dysphoria is morphemically divided (ModL<Gr dysphoria, to dys--, dis--an pherein, to bear1]" (Webster's, 1992 : 424) and has its psychological tenor scientifically defined as "A generalized feeling of ill-being, esp., an abnormal feeling of anxiety, discomfort, physical discomfort, etc." (1992 : 424). These symptomatologies are certainly frightful and result from fear of impending peril and/or intense dissatisfaction with one's social status. This illness resembles neurasthenia.

Psychosomatics

Psychosomatics are a branch of psychology and medical science that studies the interrelationships among the mindset or emotional makeup and the body. It deals explicitly "with the relationship of psychic conflict to somatic symptomatology" (Webster's, 1992 : 951).

Physical manifestations include ulcers, headaches, and anxiety-related symptoms such as chest pain and intestinal disorders (e.g. colitis).

Type A and Type C patients should be treated both psychiatrically and symptomatologically in order to obviate the deleteriousness of these dispositions.

Down's Syndrome

A congenital malady, Down's Syndrome is genetically determined and etiologically identifiable. "[It] is characterized by moderate to severe mental deficiency, by slanting eyes, a broad skull, and short fingers" (1992 : 379).

There seems to be a trisomy (triploid set, not the usual diploid of chromosomes) in the chromosome numbered 21 in man.

Mental retardation is unfortunate for all considered, yet many concerned, loving and altruistic people find reward and fulfillment working in institutions and day care centers.

Lycanthropy

This unusual mental illness in which one believes he is transformed into a wolf also has its roots in folklore. The middle ages were conspicuous for their eerie and animistic explication of natural processes and other phenomena, such as flickering gases in a marsh.

Gestalt Psychology

Gestalt psychology, a relatively modern school of psychology, studies, interprets and remedies the individual's perceptual and behavioral response, from its reaction to gestalten wholes (configurations of physical, biological, or psychological phenomena; it constitutes a functional unit not derivable by summation of its parts) particularly "on the uniformity of psychological processes," to a rejection of events such as "stimulus, percept (sense-datum), and response" (Webster's 1997 : 515).

Gestalt psyche, broadly defined as the psychology of the here and now, is endorsed and embraced by gestaltists as a viable and appropriate therapy.

Ambiversions

A psychological condition, ambiversion is indicated by elements of both introversion and extroversion.

In determination of an etiology one can conclude that the individual in question was rudimentarily praised and faulted and that this emotional, psychological, spiritual antipode in behavior is a manifestation of an approval-seeking and attention-seeking proclivity.

Instead of consigning a dual personality, an ascription of mood or temperament may variably apply and distill any detrimental implications.

Physiological Psychology

A branch of psychology, physiological psychology, "deals with normal and pathological (abnormal) factors" (1997 : 888) and how they affect an individual's psychological tenor.

Certainly one afflicted with cancer will exhibit an emotionally reactive set of symptoms and may be profoundly affected to the point of panic and hysteria.

Treatment for this reactivity should not be limited to therapy. It may require medication.

Religious Experience

Religious convictions are often cited by psychiatrists as delusional and fantastic.

Many patients claim to hear "God's voice," or believe that they are God, and thus their behavior may seem unusual and occasion a hospital stay.

Although not a criterion for whether it meets a specific standard in the DSM, it is a source of controversy and somewhat undetermined at present.

Criminal Behavior and Recidivism

Criminal behavior falls into different categories. The violent aggressor seems to evidence no remorse. These criminals are liable to kill at will. Sometimes they approach a potential victim and commit a senseless murder or through subterfuge they induce a victim to participate in some activity, often testing their own deceitful and insidious ingenuity, as does the serial killer.

The mass murderer is less crafty and tends to implicate himself, e.g. by killing himself as a matter of martyrdom. He usually has a motive.

Many crimes are nonviolent and those convicted of theft or drug dealing should not be consigned to general prison populations.

Many crimes are committed because of mental illness although the ratio is about the same.

The recidivist frequents both jails and mental hospitals and is a source of alarm because he tends to be a repeat offender.

Some theory concerning violent behavior is xyy chromosomal dysfunction. There is a high collateral rate of violence with this condition.

Hallucinations

Auditory are the most common. The individual tends to believe he is psychic, that he is reading minds and that he can insert thoughts in the minds of others. Usual gratuitous perceptions such as a running commentary or the radio directing a campaign around one are common indicia.

These fantastic powers can seem real to the individual although he may realize the irrationality of it.

Visual hallucinations are experienced also: a high speed chase among two vehicles accompanied by a loud revving engine noise, or having a conversation with a specter. The communication seems real

Often a drug-induced psychosis is attributable. The individual is doing speed or is flashing from LSD. Hallucination of gustation, olfaction, and sensations are also manifested, but are not as common.

Neuraleptics

There are four classes of medications in treating mental illness. The primary drugs are the antipsychotics and are usually administered in state-run institutions. Drugs are grouped, as for example Phenothiazines and Thioxanthenes; some of these are Prolixin, Navane, Trilafon, Stelazine, Haldol, and Clozaril.

Antidepressants or mood elevators include: cinaquon, Prozac, and Elavil.

Benzodiazepines: Ativan and Valium.

Lithium, lithium carbonate.

The fundamental treating tenet is restoration of a chemical imbalance. In theory the neurotransmitter dopamine is misfiring synaptically.

Abulia

The psychological lexicology of this word is defined as the "loss of the ability to exercise willpower and make decisions" (Webster's, 1992 : 6). A succinct and common sense consideration or practical understanding might render this individual as timid, lacking self-esteem, and/or wavering from passivity to aggression. Abulia is typically a psychic or personality deficit albeit deemed by professionals as an indication of a deeper psychological problem or conflict and can be a precursor or manifestation of a serious personality breakdown.

Manic-Depressive Psychosis

We all have mood swings, periods of elation as well as gloominess. The manic-depressive individual, however, experiences excitement, grandiosity, poor sense of judgment, and even irritability during a stretch of the manic phase of illness, with or without a period of interposition.

And almost diametrically opposed to this manic component is a depressed and dreary phase, individuated by self-disparaging and self-nullifying behavior and ideations. This depressed state is more cause for concern.

The "snap out of it" or "bring yourself together" coaxing does not work. The depressed person is in real danger of suicide. Lithium is routinely prescribed.

It's a High Tech World

In a world of scientific, technological and educational advancement, new discoveries, solutions to enigmas, and expedients to ameliorate the quality of life are being revealed at a hectic pace.

Disclosures on bacterial virulence, genetic engineering, treatments for cancer and cardiac disease, disease-detecting imaging techniques, and state of the art and futuristic equipment allows us to live longer, healthier, and happier. E. Fuller Torrey seems to have linked schizophrenia to a prenatal virus in the second trimester.

Findings in the field of genetics and their progeny are in the limelight these days. Medical advancements in research, prevention, and treatment will someday allow us to overcome just about any disease, disorder, or affliction.

The test-tube births, cloning, and cryogenics suggested by Aldous Huxley's futuristic book "Brave New World" might soon be common procedure. Similar advances are being made in the field of psychology as well.

Educational Psychology

Concerned with educational and healthy psychological development, educational psychology is expressed in human terms through maturation, aptitude, and academic achievement. Teaching through better methodologies improves guidance and direction and progress is evaluated through standardized testing (Webster's : 1992).

Our schools as well as the workplace and family units are venues of social, psychological, and intellectual development and social engineering is put to the test.

Paranoid Schizophrenia

A common mental disorder, paranoid schizophrenia is marked "by persecutory or grandiose" ideations and is manifested by delusions of jealousy and hallucinatory bouts (1992 : 854). This malady also reveals itself by suspiciousness and megalomania (feelings of omnipotence).

The paradigmatic paranoid individual is overtly, descriptively fearful and can be treated with psychotropics. Prolixin is remedial. So is Haldol, although Haloperidol has a broader range of side effects.

There is usually some incident that fashions an episode. An example is an unavoidable car accident for which the party involved nonetheless feels responsibility or guilt. Another example is the recluse, the hermit. Often this person will lose his hold on reality and think that he is being watched by animistic beings or being tracked by UFOs.

Associationism

"A reductionist school of psychology," associationists maintain that the "content of consciousness" is attributable to "association and reassociation of irreducible sensory, perceptive, and memory phenomena" (1992 : 110).

This theory relates to physical laws in the realm of reductionism. Many chemists and physicists utilize these criteria in scientific explanations, research, and discovery.

Association and reassociation in intellectual functioning is a psychological process and is apposite to both.

Mythomania

In psychiatric parlance mythomania is an anomalous and tendentiously abnormal propensity to exaggerate or lie (1992 : 785). Sometimes alluded to as a pathological condition or good-natured diversion, this behavior is puerile, self-deprecatory, and injurious.

One may find this excursion into the land of caprice and fabrication contributing to a self-eradicating series of lies covering up more lies.

I think this condition occurs when a child has an extremely fastidious and caviling parent.

Sadism

Named after the French writer, the Marquis de Sade, sadism has risen from the cimmerian recesses and has earned an opprobrious bookmark in the psychiatric annals.

Sadists can be soldiers, surgeons, sexual partners, and virtually anyone. Despite the brutal and mean denotations, some individuals have a sado-masochistic bent and find gratification in abusing or incurring abuse especially in light of their sexual fantasizing.

Inhalants

Mental hospitals and developmental centers are crammed with patients who have damaged their minds by sniffing glue, inhaling aerosol products, and paint.

Unfortunately these individuals are plagued by a dismal prospect; they will never quite regain adequate cerebral functioning.

Part of the problem, I feel, is that there is not sufficient literature to alert and forfend potential victims of this tragic problem.

Cocaine

Sometimes called the rich man's drug, cocaine, whether smoked, injected, or snorted, is an expensive habit.

The effects of this drug are euphoria, a rush of energy, and a crystal clear perception. Although pure cocaine is deadly, the conventional admixture is safer.

Some adverse effects of this drug are depression, irritability, languor, suppressed appetite, and dizziness. There are some deleterious physiological problems contingent as well as psychological ones.

Autism

A form of mental retardation, this peculiar and baffling illness has been the subject and concern of the psychiatric community and a bewildering mystery since its determination some forty years ago.

This malady, with its socially phobic and masochistic type indications, has improved with the brain molecule oxytocin. Children were observed to be more social and happier. The "autistic savant," like the idiot savant, has an ingenious ability in some intellectual areas.

Catatonia

Catatonic schizophrenia, a "psychomotor disturbance" can manifest itself in rigidity, negativism, mutism, stupor, senseless excitement or unusual posturing (Webster's, 1992 : 214). This disorder usually happens after some kind of shock or trauma, and lasts a relatively short period of time especially when treated.

Concomitant with these episodes is the sufferer's racing thoughts and a sense of peril.

Schizophrenia

Schizophrenia or dementia praecox is a broad term in defining mental illness. Initially the schizophrene experiences psychoses, personality deterioration, and a loss of contact with reality (the environment). Although rates of prevalence vary from society to society, the frequency remains at about one percent.

Here is a classic case scenario. A student at a high school is rather removed and rarely communicates with other students. He appears intelligent yet unconfident.

One day between classes he yells out loud, throws his books on the floor, and recoils into a ball in front of his peers.

One's immediate impression is that he is reacting to his social isolation. This is factual, yet not the whole story.

This high school student and preponderantly all schizophrenics live a life of emotional, psychological, and social torment. When the time comes to mature emotionally and fend or make a life for themselves, they are at a deficit. This is often the case with those afflicted with mental illness. When the time comes to leave home and assume responsibility these individuals run into problems.

Obsessions

An obsession is the term for thoughts, feelings and impulses that are an object of preoccupation or more generally understood as a continual disturbing concern with something -- physical appearance, for example. It is related to obsessive-compulsive neurosis, "an anxiety disorder in which a person feels trapped in repetitive thoughts and ritual behaviors designed to reduce anxiety" (Wade & Tavris, 1993 : 576). This behavior, usually manifested in light of some irrational fear, can again preclude sound psychological and emotional development and is usually arrogated as a modus to avoid a potentially parlous yet necessary stage of development such as working through fear. Occasionally one learns of a celebrated scientist or other personality with some idiosyncracy such as lining up his bed precisely north-south. Some of these behaviors are unusual, others preternatural.

Some rituals include hand-washing, combing hair constantly, avoiding lines on a walkway, and cutting corners off furniture. One patient I observed was so meticulous it required him twenty minutes to walk twenty feet. Obsessive- compulsive disorders are difficult to rectify, but usually pass in time.

Perversions

A perversion, an "aberrant sexual practice" (Webster's, 1992 : 878) is a condition counter to normal healthy expression. It comprises any act from masturbation to rape and fetishism.

The apparent and underlying justification to these behaviors is sexual gratification. The rapist derives a sense of power and pleasure from the egregious and violent act. One would conclude that the perpetrator experienced some form of maternal domination and this deed is a direct assault on the female order.

The fetishist can find gratification in normal expression although the object of predilection is an obsessive and irrational fixation with conventional normality. Pedophilia, a sexual perversion, may rudimentarily have its origin in the experience of two children. The pedophile either fantasizes about an activity or unfortunately and unlawfully engages in sex with a young person.

Other paraphalias are:

"Coprophilia: sexual gratification from handling feces.

Frotteurism: sexual gratification achieved by rubbing against or fondling an unsuspecting person.

Klismaphilia: sexual arousal by having another administer an enema to them.

Necrophilia: sexual intimacy with a corpse.

Telephone scatologia: gratification by making obscene phone calls.

Zoophilia: intimacy with animals" (Davison & Neale, 1996 : 348).

Psychosexuality

A term meaning a state of mind and how it relates to sexuality. Healthy psychic growth and well-being depends attitudinally, physiologically, and passionately upon the pleasures, appetency, etc., derived from this act, its fulfillment, and its relative frequency.

Good sexual expression relieves stress, promotes happiness, and is virtually essential. Relationships are steadfastly determined, for the most part, by sexual praxis.

Living in the Past

Many of us ruminate about previous experiences and good times, and have vivid and precise memories. What would nostalgia be if we did not conjure up, reminisce about, and long for past events? Sometimes, however, recalling these "good times" can materialize into an excessively impelling and obsessive preoccupation. There is no future for these individuals and these memories can interfere with or vitiate normal living, leading to melancholy, and inhibiting healthy psychological and emotional growth.

I think that these recollections develop into unhealthy patterns when we allow them. Part of this unhealthiness produces the behaviors that engage this debility. Some of these are apathy, passivity, and despair. Looking forward gives us a sense of hope, joy, and raison d'etre, and enables us to effloresce into better people.

Imago

Definitively the term imago, from the Latin for image, means "an idealized mental image of another of the self" (Webster's, 1992 : 600).

This very appropriately translates itself in Oedipus, in ego development, ego integrity, and forming enduring foundations.

Essentially this concept is important in the child's sense of deference, respect and valuation for parents, teachers, elders, and authority.

Dreams and their Significance

Many psychologists and other professionals have striven to analyze and expatiate upon the meaning and attribution of dreams, many whom have agreed with their forerunners that dreams are an emotional outlet, that dreaming released fears, desires, and fantasies, and that their interpretations had a factual and realistic basis in determining one's psychological disposition.

In ancient societies, dreams, particularly a nobleman's dreams, were interpreted by even the lowest commonality, and had an important bearing on the relative projected prosperity, token future of their kingdoms and a cogent implement in administering many affairs from polity, religion and revenue to absolute war.

Spontaneity v. Sheltering

Spontaneously-guided people find pleasant and satisfying rewards in undertaking extrinsic and exploratory excursions by venturing out of the personal microcosms and "grabbing life by the horns." These individuals are not limited by fortuitous contingencies and tend to take chances in discriminating the beneficial aspect of being in charge of one's life.

The sheltered individual for various reasons, notwithstanding, believes and is guided by the supposition that peril is immanent. This person tends to feel uncomfortable in social situations and inadequate in any walk of life. A priori considerations may point at some initial trauma, an experience in early development such as being beaten, or from feelings of compunction.

Love: A Necessity

Erich Fromm maintained the passion love is basic in healthy emotional evolvement. It is undeniably a requisite in early childhood and is essential in establishing and maintaining relationships from there on.

Love and its manifestations are sought after, invaluable, and nurtured by those individuals and are primary in attaining happiness and healthy sexual expression.

Marijuana

Marijuana produces symptoms ranging from euphoria to mild hallucinations and acute anxiety or "paranoia". Schizophrenics have experienced relapses from moderate doses.

Short term memory loss, intellectual impairment, and carcinogenic dangers are some of the long term evils.

Even if decriminalized, pot use poses many arbitrary and psychological problems. Physicians have isolated a considerable psychological harm in continued usage, a motivational syndrome. This condition does more psychological damage to those who abuse this substance and to society than any other illegal substance. As many as one in four in this society have used marijuana at least once.

Agnosia, Amnesia, and Fugue

Patients suffering these afflictions seem to incurred some form of brain damage. Sometimes the individual can regain proper functioning or melioration to the point of their previous ability.

Agnosia is the inability "to recognize familiar objects, sounds, etc." (Webster's, 1997 : 26).

Amnesia refers to loss of memory.

Dissosiative fugue is one disorder in which the person in question "experiences total amnesia, moves and establishes a new identity" (Davison & Neale, 1996 glossary : 8) until one day he just "wakes up."

Parkinson's Disease

"A chronic and progressive nervous disease of later life," parkinsonism is identified by tremors, weakness of resting muscularity, and an unusual gait" (Webster's, 1997 : 856). After every decade we live past forty, we lose about ten percent of our dopamine-producing cells. Once 80% of these brain cells have died, a diagnosis of Parkinson's disease is often adduced" (The Life Extension Foundation, 1998 : 180).

Treated by Artane, Cogentin, many of the affected experience a euphoria. There is also a regiment of fourteen components recommended by the Life Extension Foundation, and this protocol "attempts to cover the many neurological problems, neurochemical imbalances, and hormonal deficiencies associated with PD" (1998 : 181).

This disease crosses all boundaries and is extremely debilitating. Among those with PD are Muhammad Ali and Pope John Paul II.

Primal Scream Therapy

According to Webster's Ninth New Collegiate Dictionary, primal scream therapy or primary therapy is a "psychotherapy in which the patient recalls and reenacts a particularly disturbing past experience and expresses normally repressed anger or frustration, especially through spontaneous and unrestrained screams, hysteria, or violence" (Webster's, 1992 : 934).

I think this therapy can release pent-up emotions and have a cathartic effect in those appreciably inclined toward inhibited emotional expression. Emotions are components of our being and when untapped can inflate until they burst like a balloon. They can work for us or harm us.

Decompensation

This psychiatric particular indicates the "failure of a defense mechanism to prevent a mental disorder" (Webster's, 1992 : 359). In other words, the mind is unable to obviate an emotional conflict or compromise a healthier alternative, such as negation.

Heralded as the final chapter in the onset of mental illness, there is a failure of ego defense mechanisms and one consequently succumbs to depression, catatonia, psychoses, paranoia, hallucinations, etc.

Aversion

Aversion therapy treats and rectifies an individual's bad habits and/or asocial behavior through association of noxious stimuli with the dislike for these behaviors, thus inducing proper conformity (Webster's, 1992 : 119).

Aversion therapy is implemented and utilized in behavioral modification. Lower level care and treatment facilities tend to employ these methods. Treatment malls as a neoteric therapy have a greater therapeutic effect than disciplinary measures. Operant conditioning is also a viable expedient offering a reward instead of a punishment. The ancient Romans knew this also. Their maxim, "abuent studia in mores" means practices seriously pursued pass into habits.

Folie a Deux

Literally, this French idiom means double insanity. A not-so-rare condition, especially within the confines of an institution, folie a deux is the sharing of delusional beliefs among two closely associated individuals deemed mentally ill (Webster's, 1992 : 479). A classic scenario: Two female patients, released from a hospital, decide to frequent bars searching for a living Elvis Presley who remains incognito.

Cognitive Dissonance

A state of tension occurring within the person when he or she holds two synchronous cognitions that are in conflict or are inconsistent psychologically. Sometimes a person's behavior is inconsistent with his beliefs (Wade and Tavris, 1993 : 349). For example, a person realizes that smoking tobacco is harmful but continues to smoke rationalizing the possible harm. This behavior is at dissonance with the smoker's cognition of possible danger, and the individual will try to compromise and obviate the conflict, thereby reducing pain (1993 : 349).

Abnormal Behavior: 5 Categories

Abnormal behavior and its definitions vary from one therapist, psychiatrist, psychologist, and society to another. Some of the more acceptable and recognized are as follows:

(1) Statistical deviation. Abnormal behavior is anything that deviates from "normal". Unfortunately this definition would include something destructive (such as committing murder) and something desirable or valued, such as genius in some field. If enough people are doing something, we conclude it is normal, albeit the actions of Nazi Germany during the Holocaust would not ever be remotely acceptable (Wade & Tavris, 1993: 567).

(2) Violation of cultural standards. This is another instance of abnormal behavior where an act in itself breaches societal standards. Every culture has different "norms" and one behavior may be accepted in one case, and not in another (1993 : 567). "For example, psychologists no longer consider

homosexuality an illness. Also, many psychologists now regard a lack of sexual interest a disorder" (1993 : 567).

(3) Maladaptive disorder. The relegation of any behavior of maladaptive fits in this grouping. Examples are "the drinker who cannot hold down a job or a student so anxious about school that he can't write a paper" (1993 : 568).

(4) Emotional distress. An individual who sustains some sort of emotional impairment or despair can be thought of as functioning abnormally. Although the individual crosses no social lines he is "anxious, afraid, angry, depressed or lonely" (1993 : 568). Citing a nationwide survey, one in three will exhibit this disorder according to Regier et al in 1988 (1993 : 568).

(5) The legal definition. By law, the definition of abnormality is one that rests on an individual's inability to realize the consequences of his or her actions. The terms "sanity" and "insanity" only assess a person's capacity to stand trial in legal inferences (1993 : 568).

Declining Intelligence

Various studies have been undertaken to show why we slow down intellectually. Is it age? Is it organic? Is it inevitable? Here are some plausible considerations:

"The slowing-down hypothesis" maintains that it may not be cognitive impairments but slower reaction times. Some decline was disclosed even when ample time was used to complete testing in older individuals (1993 : 503).

"The disuse hypothesis" holds that what is learned in school is gradually forgotten "the longer one is out of school" (1993 : 503). However, Raymond Cattell (1971) distinguished between fluid intelligence, or "deductive reasoning and insight into complex relations," -- and crystallized intelligence, or the mastering and "use of acquired knowledge and skills." His "research shows that both kinds of intelligence increase" during pubescence and "fluid ability slowly decreases" with age while "crystallized ability remains stable" (Wade & Tavris, 1993 : 503).

"The generation hypothesis" defines "that age-related differences in cognition" are due to factors that discern "generational differences in training, nutritional factors, and stimulation." Studies that follow the differences between, let's say, a 70-year-old and a 30-year-old may be relative to the fact that "the 70-year-old did not have as much education" or "is living a less mentally challenging life" (1993 : 503).

Other "evidence for this view finds that short-term training programs" for individuals between 60 and 80 years old produce results that are commensurate with the losses for that age group, according to Baltes, Sowarka, and Kliegl in 1989; and Willis in 1987. Ordinarily, "most older people do not lose their capacity to learn" (Wade & Tavris, 1993).

"The Biological Hypothesis" suggests that there is a physiological, illness-related, medicinal, and "normal deterioration" that manifests the decline in cognitive abilities. "This is also true of the behavior of animals in controlled environments" (Salthouse, 1989; Wade & Tavris, 1993 : 504).

Erikson's Eight Stages of Development

A psychoanalyst, Erikson recognized that there are eight stages that lead to emotional, intellectual, psychological, and social maturity. If there is an unresolved conflict within one of these stages, a crisis would unfold which if unresolved would lead to unhealthy development.

(1) The first stage applied to an infant is "the crisis of trust vs. mistrust." If the baby's needs are not met," then the child will not develop the necessary trust needed "to get along in the world."
(2) The second stage concerns a toddler and establishes "the crisis of independence vs. shame and doubt." The toddler must learn to "stand on his own two feet," according to Erikson, and do so to prevent "feeling ashamed of his behavior" and doubtful of his maturation.
(3) The third stage is that of "initiation vs. guilt." Here the child acquires "physiological and mental skills, sets goals,

and realizes talents." The peril involved is the child's "sense of guilt" in fantasizing, acquiring "newfound power", and youthful instincts.

(4) In the fourth stage, "the crisis is competence vs. inferiority." Erikson defines this stage of "a school-age child" as one of a potential worker and provider. "The child is now learning" to be responsible and creative. "Children who fail these lessons risk feeling inadequate and inferior."

(5) Stage number five, experienced during puberty, "is defined as identity vs. role confusion. It is at this juncture that one comes to terms with one's direction or identity in life. If one's plans for the future [succeed at this stage], he or she will avoid confusion and wrong decisions." The identity crisis may appear at this stage as "the major conflict of adolescence."

(6) Stage six is concerned with "the crisis of intimacy vs. isolation." If an individual values himself, he will want to "share himself with someone." Although one can be successful in isolation, one is not complete without intimacy, according to Erikson (Wade & Tavris, 1993 : 506).

(7) Stage seven is "generativity versus stagnation." Now that one is involved in a relationship, he will experience either "complacency and selfishness" or "generativity, the pleasure of creativity and renewal." Parenting is an example of this development, although there are other means to fulfill this resolution, for example, "in work and in relationships with the younger set."

(8) Stage eight represents "the final crisis . . . of ego integrity vs. despair." If one feels he has acquired "wisdom, spiritual tranquillity, acceptance" of his "role in life," then he "will not fear death but be at peace" (1993 : 507).

Human Potential Movement

This trend in psychology is defined by therapies including "group therapy, encounter therapy, primal therapy, etc. . ." It "is based mainly on Freudian and Gestalt psychology," and is directed toward "self-realization" (Webster's, 1997 : 657).

With the tendency in modern times to rectify maladjusments -- emotional, social, or otherwise, especially in a highly competitive and complicated world, there is an irrefragible need to maintain and optimize one's emotional well-being. This can determine everything else in the roller-coaster of life.

What is Sexually Normal?

Ask an individual or many individuals and their answers will vary.

Why is kissing in this society considered acceptable and in other societies considered vulgar?

What does a younger person see as attractive in a fifty-year-old or vice versa?

Why do some consider an object sexually appealing or consider the same object in any form as unappealing?

In Melanesia, for example, young men engage in sex "with other males" as a proper and necessary ritual leading to maturity (Herdt, 1984). This action in our society would be deemed as abnormal, unacceptable, and even unlawful.

Studies on these issues have shown themselves to be inadequate although "some different theories" are, summoned to help explain "sexual identity" -- and these are "genetics, cultural pressures, experiences, and opportunities" (Wade & Tavris, 1993 : 353). What do you think?

Conditions that Harm Fetal Development

I think this issue is vitally important and merits scrupulous attention in precluding any other contingencies.

Although many harmful substances can be transmitted to the unborn child, here are some dangerous and potentially deadly ills.

German Measles. Rubella in early pregnancy can "affect the fetus' eyes, ears, and most commonly causes deafness. Preventing these evils lies in vaccinating the mother up to three months before pregnancy."

X rays or other radiation or toxic chemicals can result in "abnormalities and deformities."

Sexually-transmitted diseases such as syphilis have been known to "cause retardation, blindness, and many physical infirmities." Exposure to the AIDS virus and full-blown AIDS has induced death in some cases. "Herpes can affect the fetus" but only during delivery. "A cesarean section can prevent fetal susceptibility in this instance."

Cigarettes. Pregnant women who smoke have a greatly increased "likelihood of miscarriage, premature birth, and underweight babies." This can lead to "infant sickness, Sudden Death Syndrome," and hyperactivity, and lead to behavioral problems in school.

Alcohol. Studies on pregnant women who drink steadily show that up to 30 percent of their babies may be born with Fetal Alcohol Syndrome. "Infants with FAS are smaller than normal, have smaller brains, are less coordinated, and have to some degree mental retardation" (1993 : 458).

Drugs. Morphine, cocaine, and heroin have been implicated in "interfering with proper fetal development." Many other "drugs such as antibiotics, antihistamines, acne medications, diet pills, coffee, and even excessive amounts of vitamins" can be "transmitted to the fetus." Caution too should be observed in ingesting prescribed medications where studies are not conclusive in validating their safety. During "the sixties women who used the tranquilizer Thalidomide gave birth to babies with missing or deformed limbs" (Wade & Tavris, 1993 : 459).

Mental Abilities

Differences in intellectual abilities can be measured by various "techniques of psychological assessment" (1993 : 381). "Achievement tests, measuring acquired skills and knowledge, and aptitude tests, that measure the ability to acquire skills and knowledge," differ from one another in "degree and intended use," and are valid and practical.

Standardized psychological tests can be applied and scored according to established norms. Their reliability is evaluated by consistent scores and their validity (measuring what is intended to be measured) is also important (1993 : 382).

"Among psychologists and educators, however, there is a trend" that questions "the validity of even some widely-used tests." Some feel that the SAT does no more to justify predicting college performance than high school grades (1993 : 383).

The intelligence quotient is a measure of how some people have scored intellectually vis a vis others. Alfred Binet designed and organized the first widely-used intelligence test. Although it was believed these tests were measuring an inherited capacity that could not be changed, there are critics who argue that these tests should be dispensed with (1993 : 386).

In contrast with these psychometric approaches to intelligence testing, are others such as the strategies to determine an individual's ability to solve problems. "Sternberg's triarchic theory of intelligence (1988) discriminates three aspects of intelligence outside of basic IQ. These are identified as componential, experiential, and contextual intelligence" (Wade & Tavris, 1993 : 392).

New tests of conceptualization like this are being employed and improved upon, although the IQ test is the standard ruler of intelligence measurement.

Differences in mental abilities are being assessed and determined by heredity and environmental factors; this is known as the nature-nurture influence. Through these complex interactions, behavioral geneticists have been able to determine the heritability factor of IQ tests from studies with twins and adopted children.

Heritability in relation to the extent of the differences of a trait in a group of individuals is accounted for by genetic variation (1993 : 394-96).

"Heritability estimates" only apply to dissimilarities within a particular group, living in a particular environment and not between groups. Also, a highly heritable trait may be affected by environmental modification (1993 : 396).

Studies of behavior-genetics in twins and adopted children have yielded a wide range of heritability estimates for intelligence measured by IQ testing, but these estimates averaged between .47 and .58 by genetic factors. "That is, between 47 and 58 percent of the variance in IQ scores is explainable by genetic factors" (1993 : 398).

"Mental abilities can be affected by nutrition, exposure to toxins, mental stimulation, family size, individual experiences, stressful circumstances, life and parent-child interactions" (1993 : 402). Cultural differences in raising children may explain some inconsistent intelligence performances among black and white in this country and children in Asian countries and children in this country

(1993 : 405). However, these scores may not reflect other abilities that may be more valuable to that particular environment that are not measured by school testing.

May it be specified that education and social background have been determined a more valuable asset in occupational success than IQ per se. Unspectacular IQ scores may not be a key ingredient in the employment sector. Some so-called unintelligent employees may have a great deal of valuable practical capability revealed in everyday activities. Emotional intelligence is important as are drive, determination, and resourcefulness. We all have something to offer.

Memory: Functions and Stages

A classic prescription for memory would be "the capacity to store and retrieve information." Subsumed is the reference of mental structure or one that accounts for this capacity, and the material that is retained. Memory is a complex network of "abilities, processes, and mental functions" (Wade & Tavris, 1993 : 239).

Memory and some of its different functions are as follows:

Recognition: the function of identifying previously engaged materials.

Recall: the capacity "to retrieve and reproduce from memory" information and material.

Explicit memory: "the conscious and intentional recollection of a particular event or example of memory.

Implicit memory: "an unconscious retention," which can be "evidenced by the effects of a previous experience" or encountered "information on thoughts or actions" (1993 : 240).

Encoding (in memory): a process whereby information is converted into a form that can be both stored and retrieved" (1993 : 241).

Memory Systems

Sensory memory: a system that "preserves", after encoding, the retention of momentary images. There is little time utilized here, maybe a moment or so, until these images can be further processed.

Short Term Memory (STM): a "memory system involved with the retention of" images, "information for brief periods" temporally. This system "is used to store" newly perceived information and information" gathered "from long-term memory for temporary use."

Long Term Memory (LTM): permanent storage of information. Theoretically this memory has unlimited capacity (1993 : 243).

Memories We Reconstruct

Some systems of memory we reconstruct at recall along with information that we alter as it is being stored are:

Procedural Memories. Memory processed "for the performance of certain skills or actions that manifest themselves in 'knowing how'."

Declarative Memories. These include "semantic and episodic memories" and "memories of facts, data, rules, concepts, events or 'knowing that'."

Semantic Memories. These "reveal general knowledge" and are concerned with "facts, rules, concepts, and propositions."

Episodic Memories: These are for personal experiences of "events and the contexts of how they occurred."

Cognitive schema: This is "an integrated network of knowledge, beliefs, and experiences concerning a particular topic" (Wade & Tavris, 1993 : 253).

Memory and Brain Activity

Some changes in brain activity are considered to account for memory. "Short-term retention seems not to involve permanent structural changes" (1993 : 257). Long-term memory does seem to, notwithstanding. Retention for the short term involves the increase or decrease of the neuron's readiness to release neurotransmitter molecules. Studies on long-term memory in rats indicate "more dendritic growth and more synaptic connections" in acquiring motor skills and retaining them as compared to "inactive rats", as studied by Black et al. in 1990. Stimulation of various brain regions, especially the hippocampus in animals by high-frequency electrical means, mimics the activity and leads to a long-lasting increase in the strength of synaptic responsiveness (McNaughton & Morris, 1987; Teyler & DiScenna, 1987, cited in Wade & Tavris, 1993).

This in humans is thought of as a longer term process of memory and is defined as "long-term potentiation", which "is a long-lasting increase in the strength of synaptic responsiveness, thought to be a biological mechanism" and is manifested as "certain synaptic pathways become more excitable" (Wade & Tavris, 1993 : 257).

Presentation

This thought process is lexicographically presented in terms of a philosophical and/or psychological nuance. It is simply "anything present in the conscious state" at the moment in question. It is manifested "as an actual sensation" or psychologically "a mental image" (Webster's, 1997 : 1065).

Are these tendencies intermittent, persistent, infrequent, healthy? Do they interfere in terms of functioning? Do they have a pleasant or familiar quality? Can they be dreadful?

Convergence

This particular cogitational process is concerned with zeroing in on an answer or solution to a problem "by applying one's knowledge and reasoning" (Wade & Tavris, 1993 : 294).

In terms of the bicameral mind this individual is left-brain dominant. These persons appear to rationalize through logical processes, reason with linear modi operandi, and are anal-erotic. Convergents are less emotional, more aggressive as well. They are less creative, less able to think abstractly than divergents.

Reality Principle

In psychoanalysis this is the adapting of an individual's cerebral activity in order to comply with the ever-present demands of maturing and being responsible. An example is deferring pleasure (Webster's,

1997 : 1118). In life there are a myriad of responsibilities. And especially as one matures and ages, these responsibilities aggrandize and become more pronounced. Implicit here is an arrogation of some sort of a socially appropriate individualism that makes us unique. This ideography is unavoidable and has its positive/negative implications: positive in the sense that we develop, grow, mature; negative in the sense that life takes its toll.

Divergence

A right-brain process, divergent thinking employs "mental exploration of unusual and/or unconventional alternatives in problem solving" (Wade & Tavris, 1993 : 294). These functions tend to enhance creativity and abstract abilities, and can embody a fantastic or fanciful ideational orientation.

Artisans, writers, painters, and poets are some of this set. Divergents tend to be liberal politically and oral-erotic in Freudian sexual theory.

Aversion

This word in psychological parlance means "designating or having to do with conditioning, therapy, etc. . . . that produces or is intended to produce an aversion to a certain kind of objectionable or unwanted behavior" (Webster's, 1997 : 95).

Many alcoholics are functioning alcoholics, meaning their lives are not out of control. They are employed expressly for the acquisition of alcohol and they function within appropriate social parameters.

Through therapy, 12-step meetings, or self-induced disgust, they can realize this losing mode of existence and overcome their addiction.

James-Lange Theory

This theory was proposed by these two pioneers and is substantially validated as a true method of psychological-cum-physiological functioning.

It follows "that emotional amenability results from the perception of one's own bodily reactions" (Wade & Tavris, 1993 : 313). According to this view, emotions are physiologically distinguishable. An instance here, e.g., is one feeling joy as a condition of love, or pain from a rejection of affection resulting in a disturbing emotional reaction.

Hedontics

This field of scientific, psychological investigation is involved with a person's happy [and] pleasant or unhappy [and] unpleasant feelings" (Webster's, 1997 : 625). The leaders of a fledgling nation considered the "pursuit of happiness" an undeniable right. The Desiderata, an impeccable and brilliant formula for living, asserted that happiness should be the ultimate aspiration.

When one is unhappy life is dreadful. Thanatopsis is common and treatment is appropriate and necessary. Happiness is healthiness.

Humanism

Humanism is "an approach to psychology that emphasizes" the individual's "personal growth and the achievement of potential rather than scientific awareness, prediction, and controlling of behavior" (Wade & Tavris, 1993 : 16).

Apparently this individualistic and humane method toward a more nurturing mode of treatment is conducive to health and assures a successful voyage to the land of entelechy, notwithstanding severe personality defects.

Double Bind

A double bind is a psychological dilemma in which "an unusually dependent person receives conflicting interpersonal messages from a primary source, or experiences disparagement no matter what his response to a situation" (Davison & Neale, 1996, Glossary : 8). The person disparaged consequently develops complexional inferiority, ambivalence, and low tolerance for frustration. He should seek professional help and may develop schizophrenia.

Linguistic Relativity Theory

According to Benjamin Lee Whorf, our habits of thought and perception are fashioned by language and communities of different languages have differing conceptions of reality (Wade and Tavris, 1993 : 302).

The British, for example, speak basically the same language, albeit their society, culture, and values contrast with ours. Therefore there is greater disparity with languages among varying and distinct communities. Among other cultures, cross racial comparisons would differ greatly and non-racial comparisons less.

Stimulus Discrimination and Generalization

"The tendency to respond differently to two or more correlative stimuli," that differ somewhat is known definitively as stimulus discrimination (1993 : 205). S.D. "in classical conditioning occurs when a stimulus" similar to the conditioned stimulus fails to "elicit the conditioned response" from the organism. In operant conditioning it occurs when an organism responds to "the presence of one stimulus but not in the presence of other similar stimuli" (1993 : 216).

This process is an extension of classic Pavlovian psychology.

Stimulus Generalization

The basic tenet of this theory is that after conditioning, the response to similar stimuli resemble the response of "one involved in classical conditioning" (1993 : 205).

In operant conditioning, the response to stimuli occurs from either reinforced or punished conditioning when the reaction to one stimulus tends to occur or be suppressed in reaction to other similar stimuli (1993 : 215).

Thematic Apperception Test (TAT)

This is a projective personality test usually administered by a psychologist and it asks the subject to articulate or interpret a series of ambiguous scenes.

Developed by Christiana Morgan and Henry Murray in the 1930s, the TAT monitors a person's behavior and the strength of the internal motive as captured in the fantasies described in detail in response to these scenes (1993 : 366).

The internal motive is what pushes us to succeed. Why do some of us have fewer abilities but have the drive to accomplish just about anything? Why do some of us settle for less? This will be addressed next.

The Need for Achievement

In the early 1950s David McClelland determined that some "have an internal need for achievement (nAch)." This need motivates individuals "much as hunger motivates us to eat." McClelland speculated about the "psychic x-ray" that would enable us "to observe what goes on in a person's head" (1993 : 366).

People who score high on the TAT tend to set their standards high and have "achievement motivation." The question is, why does one want to achieve? My answer is basic and simple: because of reward, whether monetary, intrinsic, or approbatory.

There is also involved in explanation two factors or conditions in assessing this ideal striving. One is the "implicit (unconscious) motive" expressed by deriving pleasure in achievement itself sustained over time. The other is the "explicity (self-aware) motive" and it "predicts how one will behave in a certain situation because of immediate incentives and rewards" (1993 : 367). These motivations are unrelated to ability, and success is predicted by the goals that are set by these individuals.

Sensation

Sensation is the process for "the detection and encoding of physical energy" that is determined by adscititious or intrinsic events.

The detection of these changes are performed by sense receptors "located in sense organs" such as "the eyes, ears, tongue, nose, skin, and body tissues." The receptors involved "are structural extensions or dendrites of sensory neurons."

The various receptors are different cells with different functions "separated from sensory neurons by synapses" (1993 : 151-52). The results are "immediate awareness of sound, color, form, etc."

Sensation itself permits us to laugh, cry, or think although it is only "the handmaiden of perceptions." Perception defines objects, allows us to distinguish foods, and appreciate music, for example. It makes meaningful patterns out of sensory experience.

"Vision produces a two-dimensional image" on the retina; hearing, different notes in the ear. The former perceives a three-dimensional world, and the latter enables us to hear a three-note chord.

Sensation and perception as fields of ongoing scientific study are important mainly because they are "the foundations of learning, thinking, and behavior." These functions (S & P) are also instrumental in educating the scientific community in advancing technological strides in areas such as flight control in space craft to cybernetics and in mundane industry, from hearing aids to radios and television sets (1993 : 152).

The Object-Relations School

"Developed in Great Britain by Melanie Klein, W.R.D. Fairbairn, and D.W. Winnicott, "the object-relations school is a contemporary challenge to classical Freudian theory." At variance with the Oedipal theory of development, this school upholds the importance of the first two years of a child's life and instead of the powerful father seen in Freudianism, the relationship with the mother is crucial. This theory also maintains that the human drive is not impulse gratification but the need for relationships.

The word object connotes not the infant's attachment to a person but the nature of the "infant's perception of this person." "As the child interjects a representation of the mother," for example, as "kind or fierce," loving or domineering, it is not literally the woman herself. Whereas Freudian dogma considered the female as the problem, object-relations blames the male development. The view is that "both sexes identify primarily with the mother," although girls need not separate from the mother as males must. As a result, males are handed down an identity precarious and insecure from the female's, and the typical dilemma "for women is how to develop autonomy," and for males, "how to develop attachments" (1993 : 438).

Freud's Dissenters

Although Freud was honored as the psychiatrist emeritus, his theories innovative and himself a humanitarian making invaluable contributions to society, many have abandoned his particular tenet and have developed their own theories and proposals.

Carl Jung (1875-1961) "differed with Freud on the unconscious" and stipulated the "individual's personal unconscious." He also stated "that there is a collective unconscious containing the universal memories and history of mankind." Jung also maintained that we all share common archetypes (patterns which symbolize anything from a god to nature, a parent to a wicked witch). His "persona and shadow" are powerful archetypes we share as well; "the persona, the public personality" and the shadow, our primal fear of nature. The shadow represents the aggressive, animalistic side of humankind.

Jung recognized, like Freud, that humans share both "masculine and feminine qualities." He proposed that there are inherent in our personalities the characteristics of anima and animus. "Anima represents the feminine archetype in man," and "animus the masculine archetypes in females." Conflicts, Jung elaborated, can manifest when one denies this other feminine or masculine side.

"Although Jung shared Freud's fascination with the unconscious, he also emphasized the positive and forward-moving drives of the ego," and identified the traits of "introversion and extroversion as basic personality orientations." His research on this matter was promulgated decades ahead of time (1993 : 436).

Karen Horney (1885 - 1952) agreed with Freud on specific issues such "as sexual or aggressive motivations, genetically-based instincts, and the inevitability of psychic conflicts," but differed from Freud on many others. Horney promoted the "concept of basic anxiety," that we all feel "isolated and helpless in a potentially hostile world" (Wade & Tavris, 1993 : 436). She suggested that "anxious children compensate for this by becoming either aggressive," "overtly submissive," or they "may become selfish and self-pitying as a way of getting attention or sympathy."

Generally people behave "in one of three ways," Horney suggested. "They either move toward others, seeking love and support"; they "move away, trying to become independent;" or "they can move against others becoming competitive and critical" of others. She maintained "healthy people balance three behaviors" and that some people are too this or too that.

Horney also repudiated Freud's "penis-envy" and attitude of "female inferiority," stating that men have "womb envy" and thus "glorify their genitals" because they cannot give birth" (1993 : 437).

Alfred Adler (1870-1937) differed with Freud and his unconscious drive of the id; its aggressive, self-gratifying and primitive aspect. Adler felt "that people have within them a drive for superiority," and "not to dominate others but to strive for self-improvement."

He wrote that certain individuals "develop an inferiority complex" by "pretending to be strong and capable," in denying "their natural limitations" and that they "become overly concerned with protecting their self-esteem instead of coping with real problems."

Adler emphasized the individual's dependence on others, "their empathy and concern for others" in "that of social interest." This basic striving predicted "the ability to be unselfish, sympathetic, and cooperative."

Adler also stressed that "the human personality" in "essence is the creative self," forming "his or her own personality from heredity and experience," so that "all psychological processes form a consistent organization within everyone."

This "personality structure is thus expressed in a unique style of life," "a result of biological factors," "conscious and unconscious motives, history, and life goals" (Wade & Tavris, 1993 : 437).

Love

An elusive yet attainable commodity, love throughout the ages has been the subject of poets, the desire of youth, and the primordial object; the want and striving from warriors to the poor, intellectuals to the aristocracy.

The Greeks personified love in the mythopoeic figure of Aphrodite and the love-object was a constant source of Greco-philosophical expression as in many other societies.

Passionate v. Companionate Love

This rendering "contrasts passionate love with its characterized turmoil of emotions with companionate love" and its proper "affection and trust." "Passionate love is intense," focused, and sexual; often "unstable and fragile" while "companionate love is calm," also focused, more "stable and reliable." Passionate love manifests itself in "crushes, infatuations, love at first sight, and love affairs," transitory and not enduring.

Some professionals feel that passionate love is a biological process "rooted in our evolution," that renders us bonded as couples, reproducing organisms, and "looking after each other" (Wade & Tavris, 1993 : 361).

Six Styles of Love

John Allen Lee (1973, 1988, cited in Wade & Tavris, 1993) proposed six different "styles of loving" identified with Greek names. He believed there to be three basic ones:

"(1) Ludus (game-playing love).
(2) Eros (romantic, passionate love)

(3) Storge (affectionate, friendly love)"

He postulated three secondary styles as well:

"(4) Mania (possessive, dependent, 'crazy' love)
(5) Pragma (logical, pragmatic love)
(6) Agape (unselfish, brotherly love)

Some individuals are motivated toward affectionate, friendly love; others are more playful. Some types are more pragmatic; others have a lower standard and still others have a candid disposition.

The Attachment Theory of Love

This type of love is characterized by "physical connection." Cindy Hazen and Philip Shaver (1987) believe that three kinds of infant attachment -- secure, anxious/ambivalent, and avoidant -- determine "adult styles of love." Their research concludes that "some couples are strongly attached; they are not jealous, worried, nervous, or concerned about abandonment." Other adults are fretful, "anxious, and ambivalent;" they are afraid "their partners will leave them." Still others "distrust and avoid all attachments" (Shaver, Hazan and Bradshaw, 1988, cited in Wade & Tavris, 1993 : 362).
There is overlapping here with these definitions. Theories discern "that most people can and do change their styles of love over time and with new partners." People whose first love is a "romantic and idealistic relationship" can become "more realistic and even cynical" (Carducci & McGuire, 1990, cited in Wade & Tavris, 1993). It is better to have loved and lost.

Women as Castrated Man: Sigmund Freud in Agonito

Freud explained that gender is determined by the preponderance of male and female characteristics; secondary sexual characteristics. This is the case too, he states, in mental dispositions. In sex, the male is active; the female passive, although Freud held that both sexes are bisexual.

In women this passive role in sex might carry over into strictly feminine roles. And women's suppressed aggression may develop into masochistic impulses.

In the Oedipal stage, a young girl is to change her object-cathexis (to her father) and also her erotogenic zone. Boys do neither. And here too, but later, women fantasize about being seduced by the male parent, as expressed in typical Oedipal complexes.

With the castration complex, both sexes physically distinct have a resulting psychical reaction, girls blaming their mother for lacking the male organ.

Freud further states that it is this complex in which the female expends energy (cathexis) and sublimates (a modification such as a profession or career in reaction to the expressed wish to have an organ.) This may relate to an infantile factor, but to dismiss this, is somewhat infantile.

Three developments tend to actualize: (1) sexual inhibition, (2) masculinity complex, (3) normal femininity.

Often the females' suppression of sexual instincts (masturbation) arouses sympathy for others with this dilemma and it may also be a condition for marital choice. Here the desire for a baby takes the place of the "organ".

In the boy the Oedipus stage leads to a castration complex, but in the girl the castration complex conduces to Oedipus.

Female homosexuality results from paternal displacements. Lesbians play the parts of mother and baby instead of husband and wife.

Social breeding as well as sexual functioning, contributes to narcissism in the female. They are more vain because of sexual inferiority. Shame emerges as another characteristic, a manifestation of genital deficiency.

Because a wife in her initial Oedipal stage loves her mother and a man his mother, their union might be considered a phase apart psychologically.

These beliefs -- namely, woman's social inferiority, sexual passivity, physiological inadequacy, and shrewish disposition -- are certainly considered Victorian today.

And women historically were unlikely to occupy positions of wealth, authority, and influence. Perhaps the course of nature and the equality of men and women contemporarily are actually concomitants, with neither sex being implicitly more capable or superior than the other.

Emotions

There is a medieval legend involving a knight named "Ulrich von Lichtenstein who went on a pilgrimage from Venice to Bohemia. According to lore and his own account, he challenged and broke 307 lances, unhorsed four opponents, and accomplished his five-week expedition undefeated."

The reason behind these wonderful exploits "was his passion for a certain highborn princess whom he sincerely adored."

Ulrich supposedly endured intense feelings of longing, woe, and melancholy, and a state of fondness that made him intensely prepossessed and that occasioned this amazing campaign" (Wade & Tavris, 1993 : 311).

What if there was no emotion, no love, or no attraction? How would we justify his knightly performance? This story would not have occurred.

Emotions are what make us human, give life meaning, motivate us, produce stress, aid in development, and organize personality.

Emotion and the Body

Inchoately philosophers considered emotions contingent "on mixtures of bodily 'humors': blood, phlegm, choler, and black bile." If you were hotheaded or irritable, you were thought of as "having an excess of yellow bile (choler)." "If you were slow and unemotional it was because of too much phlegm" (1993 : 312).

These assumptions were clearly farfetched but where and how does emotion proceed from our physiology? William James was one of the first pioneers to explicate this enigmatic phenomenon.

He concluded in his sequence of physico-emotional order that: (1) "something happens," (2) one reacts with certain physiological responses, and (3) this converts to emotional reactions from one's experience and interpretation of these responses.

James is quoted as saying: "We feel sorry because we cry, angry because we strike, afraid because we tremble." His theories were not apodictic, but he pioneered the way for research into "facial expression, cerebral processes, and the autonomic nervous system" (1993 : 313) for answers.

The Indications of Emotion

Darwin expressed that facial variations into emotions -- "the smile, frown, grimace" -- are all biological or inborn. This is evident because primal man needed to distinguish between a friendly stranger or one who was hostile and threatening; his ability to perceive this difference is also biologically instinctual. Modern psychologists support this assertion.

Two researchers -- Ekman and Friesen -- have gathered validation "for a universality of some six facio-emotional expressions: anger, happiness, fear, surprise, disgust, and sadness" (Ekman, Friesen & Ellsworth, 1972; Ekman et al., 1987). They've recently made an additional claim -- contempt (Ekman & Heider, 1988, cited in Wade & Tavris, 1993).

Many varying cultures can identify one another's facial gestures without the least understanding of these emotions (Wade & Tavris, 1993 : 313). It is also necessary for an infant to recognize its parent's expression as a warning or affection, as a survival mechanism (1993 : 314).

Mental Illness and Contemporary Thought -- Somatogenesis

"After the fall of the Greek and Roman empires," Galen's writing was deemed the standard and accepted source of documenting and defining illnesses. New findings did emerge during the Middle Ages, however, and Vesalius' disclosure that Galen's methods were incorrect gave a boost to more factual and modern scientific investigation.

The distinguished English physician Thomas Sydenham (1624-1689) announced an "empirical approach to classification and diagnosis" that greatly "influenced those interested in mental illness."

One such physician, Wilhelm Griesenger, a German, insisted that mental illness had a physiological cause. This view mirrored that of Hippocrates and his somatogenic espousal.

"A well-known follower was Emil Kraeplin", who published a "textbook of psychiatry" in 1883, and furnished an organic-oriented system of mental illness. He discerned a tendentious nature of symptoms that appeared together and called them a syndrome in order to understand them in light of an "underlying physical cause." He regarded distinct mental illnesses as having "a different genesis, symptomatology, course, and outcome," and proposed that schizophrenia was caused by a chemical imbalance."

Around these times a startling discovery took place in medical history, "the full nature and origin of syphilis." This disease was known to cause "both physical and mental impairment," and in 1825 it was designated a disease, general paresis. It took until the 1860s and 1870s, when "Louis Pasteur established the germ theory (infection of the body by minute organisms)" that a "link was based between infection, destruction of certain areas of the brain, and psychopathology." Because one form of psychopathology had a biological etiology, then so might others, was the thinking. Somatogenesis and the search for its credibility was off and running (Davison & Neale, 1996 : 19).

Psychogenesis

The pursuit of a somatogenic etiology was carried on well into the twentieth century. But in different European communities at the onset of "late eighteenth and throughout the nineteenth century" (1996 : 19), there was a quest for an entirely different modus in understanding mental illness. Here were "various psychogenic points of view" promulgated especially in "France and Austria."

During this time many were affected with hysteria (the modern-day term is conversion disorder). "An Austrian physician, Franz Anton Mesmer" (1734-1815) (1996 : 20), attributed this disorder to a certain "distribution of a universal magnetic fluid in an individual." And in treating these patients, he would have them "sit around a covered baquet (tub) with rods protruding through the cover with various bottles of chemicals beneath" (1996 : 21-22). He would then touch a patient with these rods. The rods were believed to impart "animal magnetism, thereby removing the hysterical anesthesias and paralyses." Because of these practices, Mesmer was "considered one of the earlier practitioners" of contemporary hypnosis; "mesmerize was the older term for hypnosis."

Another pioneer in the psychogenic movement was Jean Martin Charcot (1825-1893), a renowned Parisian neurologist. He was deceived by some of his students who proceeded to "hypnotize a normal woman who they suggested had hysterical symptomatologies." When she was awakened and consequently freed of her symptoms, Charcot was impressed and adopted a nonphysiological mindset.

In a celebrated case in Vienna, Josef Brueur (1842-1925) treated a woman dubbed Anna O. This patient "was bedridden with several hysterical symptoms; legs and right arm and side were paralyzed, her sight and hearing were affected, and she had difficulty speaking." She would also "mumble to herself, ostensibly preoccupied with disturbing thoughts."

After hypnotizing the woman, Breuer "repeated to her some of her mumbled words." She gradually began to "talk more freely and more emotionally about her troubled past." Breuer revealed "that relief and cure of these symptoms" were attained through recalling "a precipitating event for these symptomatologies," especially "if the original emotion was expressed" (Davison & Neale, 1996: 22). This came to be known as the cathartic method.

The Mental Hospital Today

Throughout today's society, with its booming metropolises, nature preserves, jet planes, and space stations, over two million of our inhabitants are hospitalized for major mental illnesses. Most of those hospitalized are in custodial care at many state facilities referred to as state hospitals.

The annual costs are exorbitant.

With patients on private insurance the care at privately-run hospitals can exceed a thousand dollars a day. Some extravagant institutions are Cedars-Sinai and McLean hospitals. However, most patients are indigent and reside in low-level care facilities.

Some hospitals specialize in what is considered treatment for the criminally ill or insane. Usually one who breaks the law and is deemed unable to stand trial is remanded to one of these, for example, the Mid-Hudson Psychiatric Center in New York State. Although it is a hospital, it is highly regimented and secure.

Some of the drawbacks with state-run facilities is that funding is inadequate. Many patients are unfortunately given too little therapy and the wards are bleak and cheerless.

There is a general feeling of hopelessness and abandonment, as in being imprisoned. Many institutionalized patients fear being discharged from these places, and regrettably want to remain in them (Davison & Neale, 1996).

In milieu therapy, however, there is a trend whereby the entire "hospital community is converted to a therapeutic environment. Social interaction and activities are geared so that group pressure directs patients toward normal functioning and they are treated more as responsible adults" (1996, 21). They in turn are allowed more freedom.

A token economy, detailed with a patient's history and "other behavioral therapy interventions" that are "tailored to a patient's individual needs" is a way of reinforcing behaviors compatible with the appropriate situations. The patient is thus rewarded with tokens for privileges or other items (1996 : 21). This has been lauded as an efficacious treatment procedure or measure.

Psychoanalytic Theory of Depression

Freud commented on the "potential for depression" established early in development in his eminent paper "Mourning and Melancholia" (1917). In it, he declared that a child's needs that are either "insufficiently or oversufficiently gratified" may effect a state whereby the child remains dependent on "instinctual gratification" circumstantial to it. In this arrest of healthy maturation, during the oral stage, a tendentious dependence on others for self-esteem may develop.

Later, as an adult, a complexity of factors contribute to depression.

After losing a loved one, Freud theorized the mourner to introject, or incorporate, this lost person. Here he or she identifies with the lost one, presumably in an otiose attempt to recover the loss. "Because," as Freud asserted, we "harbor negative feelings toward those we love" at an unconscious level, the aggrieved is now the "object of this hate and anger." Concomitantly the mourner feels abandoned and manifests guilt for contingent sins "against the deceased."

After this, a period of "mourning work" surfaces when the mourner reminisces about the lost one and thus separates the bereft from him or her, loosening the bonds imposed by the act of introjection.

However, in "overly dependent individuals, the grief work can go astray" and result in a "process of ongoing self-abuse, self-blame, and depression." These individuals fail to loosen their bonds with the loved one and proceed to punish themselves for the "fault and shortcomings" arrogated "to the loved one who has been introjected." This mourner continues to direct this anger inward and is testimony for the "psychodynamic view that depression is anger turned against oneself."

In the case of depression not incurred by the loss of a loved one, psychoanalysts maintain the concept of "symbolic loss" as a means to insure this "theoretical foundation." As an example, "a person may unconsciously consider a rejection as a complete withdrawal of love."

Much data, however, does not support this point of view. Beck and Ward (1961, in Davison & Neale, 1996), determined through dream analysis, "themes of loss and failure, not anger and hostility." Also, "projective tests established that depressed people identify with victims and not aggressors" (Weissman, Klerman, & Paykel, 1971, in Davison & Neale, 1996) found that intense anger was expressed toward others close to them instead of being internalized.

Although Freud's theories are rejected today, his "basic suppositions have a lingering influence." For example, the irrational, self-effacing statement "It is a dire necessity that I be universally loved" of which Ellis lay claim to "much human suffering" would certainly have a devastating effect on "Freud's oral personality after the loss of a loved one."

And there is "a large body of evidence" too, that depression is brought about by "stressful life events" and that these include losses, such as unemployment or divorce (e.g. Brown & Harris, 1978, in Davison & Neale, 1996: 220).

Personality Disorders

Initially called character disorders, personality disorders are a constellation of disorders characterized by persistent and inflexible examples of "inner experience and outward behavior" that conflict with "cultural expectations and cause distress or impaired functioning."

It is prudent here to cite the "medical student syndrome" that along with many descriptions of illness and disorders, we have a tendency to find similar traits in our personality and that of others. This is normal (1996 : 263). Only when one is impaired to the point of "long-standing, pervasive, dysfunctional patterns" can we justifiably attach these descriptions of disorders.

Some problems with classifying personality disorders are:

(1) The DSM has categorized these disorders "as indicated on Axis 2. This means their presence or absence are determined whenever a diagnosis is being made." When one is diagnosed, for example, with panic disorder, this is also considered in light of whether "a personality disorder is also present."
(2) "Over the years personality disorders have had" classificatory problems. Because of poor clarity and reliability in some "field trials of DSM 3", "although personality disorders were diagnosed in over 50 percent of" the cases, their reliabilities were inadequate (American Psychiatric Association, 1980, in Davison & Neale, 1996).
(3) "Problems remain with DSM 3." Often a patient diagnosed with a particular disorder also demonstrates indications of traits of many other disorders.
(4) There is an overlap problem with diagnoses. "Widiger, Frances, and Trull (1987) found that 55 percent of patients with borderline personality disorder also" displayed "diagnostic criteria for schizotypal p.d.; 47 percent met criteria for antisocial p.d.;" and "57 percent for histrionic p.d."

This is discouraging when we try to define individuals with a particular "disorder especially with some control group." If we agree "that people with borderline personality differ from normal people, have we learned anything specific to borderline personality disorder? Do the findings relate to personality disorders in general or another diagnosis? Because changes in diagnostic in DSM 4 are minor", it is presumed that the overlap has not been classified (Davison & Neale, 1996, 264).

"Personality disorders are now grouped into three clusters of diagnoses in DSM 4": cluster A (paranoid, schizoid, and schizotypal), who seem odd or eccentric; cluster B (antisocial, borderline, histrionic, and narcissistic) with their dramatic, emotional, erratic bent. Those in cluster C (the avoidant, dependent, and obsessive-compulsive) appear anxious or fearful.

Paranoid Personality Disorder

"The paranoid personality is suspicious of others, frequently angry and hostile, expects to be mistreated or exploited, becomes secretive," looks for "signs of trickery and abuse," is "reluctant to confide in others, bear[s] grudges, blame[s] others" for their faults, and is "extremely jealous" and insecure.

Such people are "preoccupied with unjustified doubts about the loyalty or trustworthiness of others." They sometimes "read hidden messages into events, for example, they may believe the neighbor's dog" is trying to "disturb them." There is a strong overlap "with borderline and avoidant disorders" (Morey, 1958, in Davison & Neale, 1996). Paranoid personality is more frequent with "first-degree relatives of patients with delusional disorder and schizophrenia. This suggests a genetic proclivity among them" (Kendler, Masterson & Davis, 1985, in Davison & Neale, 1996).

Schizoid Personality Disorder

This individual "does not desire or enjoy social relationships, has few close friends, appears dull and aloof, has no warm, tender feelings for others, rarely shows strong emotions" (1996 : 265), "are not interested in sex", or are not in want of pleasures. They are "indifferent to praise, criticism, and the sentiments of others." They are "loners who pursue solitary interests."

"The frequency of the schizoid diagnosis increased remarkably from DSM-3 to DSM-3R." Many diagnoses of "schizotypal are not diagnosed schizoid." For "overlap with other p.d.s" consistencies "are highest with avoidant (53 percent) and paranoid (47 percent)."

Schizotypal Personality Disorder

This "modern concept" of personality disorder "grew out of Danish studies of the adopted children of schizophrenics" (Kety et al., 1968 : in Davison & Neale, 1996). Although many of the "children developed full-blown schizophrenia as adults," more developed an "attenuated form of schizophrenia."

"The diagnostic criteria for schizotypal personality" were "devised by Spitzer, Endicott, and Gibbon (1979, in Davison & Neale, 1996). "Incorporated by DSM-3," they were "narrowed somewhat in DSM-3R and DSM-4."

The schizotypal personality has basically the same "interpersonal difficulties of the schizoid, excessive social anxiety that does not" lessen "with familiarity. Other eccentric symptoms," those that do not "warrant a diagnosis of schizophrenia," are also typical. Such people "may have odd beliefs or magical thinking (superstitions), beliefs that they are clairvoyant and telepathic (telepathiological) and demonstrate recurrent illusions -- they may sense the presence of a force or a person not there." Speech may be odd, "unusual, or unclear." Behavior may "be eccentric;" they may have "ideas of reference (beliefs that events have an unusual meaning for the person)."

They can have "suspicious and paranoid ideations." They have a "flat and constricted" affect. According to Widiger, Frances, and Trull (1987), paranoid ideations, ideas of reference, and illusions were most telling" (Davison & Neale, 1996).

As far as prevalence this "disorder is estimated at about 3 percent and slightly more frequent in men." Relatives are "at increased risk" (Siever et al, 1990, in Davison & Neale, 1996). They share some biological factors with schizophrenics (e.g., levels of monoamine oxidase, Baron et al, 1984, in Davison & Neale, 1996). There also seems to be a relationship "through genetic transmission of a predisposition." In studies, "more first-degree relations of schizophrenics are given the diagnosis than relatives of individuals in control groups. This disorder may thus be a mild form of schizophrenia" (Spitzer, Endicott & Gibbon, 1979, in Davison & Neale, 1996).

The problem of overlap with other personality disorders, according to Morey (1988, in Davison & Neale, 1996), are the following:

"33 percent of DSM-3R diagnosed schizotypal personalities met the diagnostic criteria for borderline personality disorder, 33 percent for narcissistic personality, 59 percent for avoidant personality disorder, 59 percent for paranoid personality, and 44 percent for schizoid personality disorder."

This lack of uniformity also appears in other areas of consideration. Squires, Wheeler et al., 1988, in Davison & Neale, 1996, "found no differences in the rate schizotypal disorder" in "children of schizophrenics and those of parents with mood disorders."

Borderline Personality Disorder

Borderline personality disorder is characterized by: "instability in relationships, mood, [and] self-image;" emotional instability; and anger. Persons with this condition "are argumentative, irritable, and sarcastic." They can be self-destructive, going on bouts of "gambling, spending, sex, and eating." They have an unclear and incoherent sense of self and "values, loyalties, and choice of career." They fear solitude and abandonment. They seem "to have a series of intense one-on-one relationships, usually strong and transient, that alternate between idealization and devaluation." They are "subject to chronic feelings of depression and emptiness," and may feign suicide. They may exhibit "paranoid ideations and dissociative symptomatologies during periods of high stress" (Davison & Neale, 1996 : 266). Most critical here are the features of "unstable and intense interpersonal relationships" (Modestin, 1987, in Davison & Neale, 1996).

Clinicians and researchers "have given the concept different meanings"; originally it "implied that the patient was "between neurosis and schizophrenia." The DSM standard "no longer has this definition."

"The DSM-3 criteria for borderline personality [disorders] were established through a study done by Spitzer, Endicott, and Gibbon (1979, cited in Davison & Neale, 1996)." The identity of "schizotypal personality disorder as a cluster of traits" were "related to schizophrenia." And "they also identified another syndrome not related to schizophrenia through a predisposition. This became the DSM-3's borderline personality disorder. Now the DSM-4 maintains these changes."

It is "apparent that borderline personalities make up a very heterogeneous group. Pope and colleagues (1983) found that about half their sample" included persons suffering from "major depression or bipolar disorder." These responded well to treatment. "Pope found too, that most of these could also be diagnosed as having another personality disorder: histrionic, narcissistic, or antisocial" (Davison & Neale, 1996).

Much overlapping "has been reported between DSM-3-diagnosed borderline and schizotypal personality disorders" (Serban, Conte, & Plutchik, 1987). This appears to have changed "with DSM-3R and DSM-4 criteria. Morey (1988) reports less overlap between schizotypal and borderline, but more overlap between borderline and histrionic, narcissistic, dependent, avoidant, and paranoid personality disorders" (Davison & Neale, 1996).

"The prevalence of borderline personality disorder is about 2 percent and is more common in women" (Swartz, et al. : 1990). There are higher incidences in cases of "childhood physical and sexual abuse" (Ogata et al., 1990), and "the disorder begins in adolescence" (McGlashan, 1983). There is a genetic disposition "with high rates in first-degree relatives of index cases" (Baron et al., 1985; Loranger, Oldham & Tulis, 1983). "Borderlines are likely to have an Axis 1 mood disorder" (Manos, Vasilopoulou & Sotorou, 1988). "Therapeutic outcomes tend to be poor" (McGlashan, 1983). "Lithium has been of some value in treating the impulsiveness" characterized by the disorder (Links et al., 1990, in Davison & Neale, 1996, 267).

Histrionic Personality Disorder

"Formerly called hysterical personality," this diagnosis "is applied to those that are overly dramatic and attention-seeking." Although these individuals "display emotional extravagance," they are emotionally shallow. "They are self-centered, concerned with physical attractiveness and uncomfortable when not the center of attention. They believe their relationships are more intimate than they are, are sexually provocative, easily influenced by others, and their speech is vague and lacking in detail."

The prevalence is equal among "men and women (2.1 percent) in a community survey" (Nestadt et al., 1990). It is more frequent "among separated and divorced people. It is associated with high rates of depression and poor physical health" (Nestadt et al., 1990, in Davison & Neale, 1996). The major overlap is with borderline personality."

Narcissistic Personality Disorder

These individuals "have a grandiose view of their uniqueness and abilities; [and] have a preoccupation with fantasies of great success." They are prodigiously self-centered, "require almost constant attention and excessive admiration, and believe they can only be understood by special or high-status people. Their relationships are disturbed by their lack of empathy; feelings of envy; arrogance; taking advantage of others; and feelings of entitlement, expecting others to do special, not-to-be-reciprocated favors for them."

"Kernberg (1970) described the grandiosity and egocentric behavior of narcissists as a defense against the rage they feel toward their parents, 'as being cold and indifferent.' Kohut (1966) proposed that this n.p. develops as a way of coping with perceived shortcomings in the self that rankle, because parents do not provide support and empathy" (Davison & Neale, 268).

"From DSM-3 to DSM-3R, the frequency of the diagnosis increased markedly. It overlaps greatly with borderline personality disorder" (Morey, 1988, in Davison & Neale, 1996).

Avoidant Personality Disorder

This diagnosis applies to individuals "who are keenly sensitive to the prospect of criticism, rejection, or disapproval. They are reluctant to enter into relationships unless they are assured of being liked. They are afraid of saying something foolish or being embarrassed by flushing or other signs of anxiety."

They feel "incompetent and inferior, and typically exaggerate the risks in doing something beyond their usual routine." Thus they "avoid work or school activities. There is substantial overlap between avoidant and dependent personality disorders" (Trull, Widiger, & Frances, 1987) "and of borderline personality disorder" (Morey, 1988, in Davison & Neale, 1996).

Dependent Personality Disorder

"The dependent personality lacks self-confidence and self-reliance." These people usually avoid responsibility by "allowing their spouses or partners" to decide "where they should live," their employment, and their friends. They have trouble "initiating anything on their own" and by "fear of losing approval" they submit to others "even when they know they are wrong. They feel uncomfortable when alone and are often preoccupied with fears of taking care of themselves. They are unable to make demands on others" and often "subordinate their needs to ensure their protective relationships they have established." They seek relationships urgently after losing one.

"The DSM-4 diagnosis contains two types of symptomatologies", one that describes "dependent behavior," and one that describes an "attachment problem" (Livesley, Shroeder, & Jackson, 1990, in Davison & Neale, 1996). Attachment "is regarded as important for personality development. The infant becomes attached to an adult and uses the adult as a base from which he or she explores and pursues other goals. Separation here leads to anger and distress. It is possible that the abnormal attachment behaviors" exhibited "in dependent personalities reflect a failure in the usual developmental process."

"Reich (1990) revealed that relatives of men with dependent personality disorders demonstrated a high rate of depression, whereas relatives of women with the disorder had a high rate of panic disorder. Dependent personality overlaps strongly with borderline and avoidant personality disorders (Morey, 1988). It is linked to several Axis 1 diagnoses as well as poor physical health" (Davison & Neale, 1996).

Obsessive-Compulsive Personality Disorder

This individual "is a perfectionist." He/she is "preoccupied with details, rules, schedules, and the like (Davison & Neale, 1996 : 269).

They are concerned with "details that projects are never finished" (1996 : 269-70). These individuals are work- rather than pleasure-oriented, have trouble making decisions and allocating time." They have problems with relationships because "everything has to be done their way. They are inflexible especially when considering moral issues." They tend to hoard "worn-out and useless objects, money, and be miserly. Obsessive-compulsive personality disorder is quite different from obsessive-compulsive disorder and do not include the obsessions and compulsions that define the latter disorder" (Davison, 1996 : 270).

Antisocial Personality Disorder (Psychopathy)

I have chosen to give a commentary on this disorder out of all personality disorders as a matter of interest (personal as well as universal). This subject is always in the news; someone commits a crime and we all want to know what happened, who did it, and what are the consequences. We seem to have a morbid interest in knowing what evil another individual has perpetrated.

Psychopathy, otherwise known as sociopathy, is the focus of novels: Sidney Sheldon's "Bloodline" for example, and movies: "The Silence of the Lambs," starring Anthony Hopkins.

Back in the 1970s, a serial killer roamed the dark streets of New York shooting lovers as they necked in parked cars.

A pathetic loser went on a spree in the 1980s butchering college women across the nation.

These are just some examples of the glorified media attention that we confer upon these sanguinary butchers.

There are more than a million criminals in jails and prisons throughout our society. Are all these individuals bad people who want to hurt others? Are these malefactors corrigible? Do all prisoners belong behind bars? What induces an individual to go as far as to murder another? What can we do about these individuals?

"In consuetude the terms antisocial personality disorder and psychopathy (and sometimes sociopathy) are used interchangeably." There are differences, however, among them. Antisocial behavior is an "aspect of both," and the history of defining "antisocial behavior is interesting" nonetheless.

In the early "19th century, Philippe Pinel conceived of manie sans delire," choosing the term in indicating a patient who "was violently insane (manie)," but did not demonstrate "symptoms (sans delire) common with the insane. In 1835 James Prichard, an English psychiatrist, delineated the disorder moral insanity in "attempting to define some behavior so morally disparate from legal codes that it seemed a form of lunacy." Prichard was referring to a man of the aristocracy, who "whipped a horse, kicked a dog to death, and threw a peasant woman into a well."

The DSM-4's definition of antisocial personality disorder has "two major components." Initially there must be a conduct disorder diagnosed before the age of fifteen, e.g., "truancy, running away from home, frequent lying, theft, arson," vandalism. The second component involves a continuation of this behavior in adulthood.

The adult antisocial personality demonstrates "irresponsible and antisocial behavior by not working consistently, breaking laws, lying, being physically aggressive, defaulting on debts, and by reckless behavior." He or she has "no regard for truth nor remorse for misdeeds."

Rates for adult Americans who are antisocial personalities are about 4 percent for men and about 1 percent for women (Robins et al., 1984, in Davison & Neale, 1996). Not only "pimps, confidence artists, murderers, and drug dealers" 'qualified' for this relegation, but so are "business executives, executives, professors, politicians, physicians, plumbers, bartenders, etc."

Henry Cleckley used the concept of psychopathy in his classic book The Mask of Sanity (1976). With "his vast clinical experience" he set down some criteria that differs from DSM's, and refers "more toward the psychopath's psychology." He indicated that "one key characteristic is a poverty of emotion, both positive and negative." According to Cleckley, "psychopaths have no sense of shame" and any "positive feelings" they have "for others is merely an act. The psychopath is superficially charming and manipulates others for personal gain." Their "lack of negative emotions make[s] it" hard for them "to learn from mistakes" and their "lack of positive emotions" may "lead them to behave irresponsibly toward others." Also Cleckley indicated that psychopaths are motivated more for thrills rather than a need, such as money.

"More researchers identify psychopaths now using a checklist developed by Hare and associates" (Hare et al, 1990, in Davison & Neale, 1996 : 271-72). "The checklist identifies two clusters of psychopathic behavior." The first cluster individuates "a selfish, remorseless individual who exploits others." The second category defines "an antisocial lifestyle." And "among axis 1 diagnoses, alcohol and other drugs are abused" (Smith & Newman, 1990, in Davison & Neale, 1996.)

It is important here to separate the concept of antisocial personality disorder from criminality, although "75 to 80 percent of convicted felons met the criteria" (Davison & Neale, 1996 : 272).

Treating Personality Disorders

There is not sufficient literature relevant to treatment protocols or effective therapies in obviating or mitigating the overt symptomatologies of personality disorders. It may be easier to turn flesh to stone, literally, than modifying the bizarre, inappropriate, and sometimes amusing demeanors that baffle medical science.

In one-on-one therapy it is important that the professional not be manipulated or fooled by the exaggerations, lies, aggressivity, etc., that the patient contrives, often as a measure of how far he or she can go. It is also necessary for the therapist to be sensitive to some of the bugbears or emotional pains with which an individual may have had a particularly distressing experience.

Medications may be efficacious in eliminating some of the perceivable manifestations (aggressivity, anger, suspiciousness, anxieties, etc.).

People, if need be, can change.

Mental Illness and How Society Reacts to the Phenomenon

Many times mental patients are the object of ridicule and social exclusion. And unfortunately this augments their feelings of worthlessness and low self-esteem, not to mention their perpetual confusion about: What can I do to be normal? How can I fit in?

As a serious social deficit, many patients develop an unhealthy constellation of mordant and indignant reactivity. Consequently this reactivity adds to further isolation and hence a bilious disposition. Thus lower tolerance for social criticism and in some cases an asocial proclivity surfaces. On the flipside there are those caring and concerned individuals who understand and sympathize with this debilitating and unfortunate condition.

Cathexis

To cathect or manifest cathexis has its rudiments in psychoanalytic theory and its implementation. Descriptively the lexicology of the words is expressed as a "concentration of psychic energy on some particular person, thing, idea, or aspect of the self" (Webster's, 1997 : 222).

This consequently has its positive and healthy, or negative and unhealthy manifestations. If an individual focuses on a psychic flaw (a compulsive disorder, for example) in order to ameliorate the condition, this is healthy psychiatric attitudinizing. Detrimental is a focus of psychic energy on conflictual ideations, such as a love/hate emotional bent with a parent, for example.

Eros

In Greek mythology, "Eros is the god of love, the son of Aphrodite, and is identified with the mythicized Roman god Cupid."

In smaller case (uncapitalized) morphology, eros is "sexual love or desire" and is synonymous with concupiscence.

In psychoanalysis and its metatheory, eros is "the life instinct, based on the libido, sublimated impulses, and self preservation" (Webster's, 1997 : 461). There is some relationship between eros and ego defense and both are tendentiously variable. Eros is innate, ego defense is acquired.

Zoanthropy

Like lycanthropy, this word has a strange and eerie connotation. The condition occurs when an individual "believes he is or can be changed into a beast" (1997 : 1555).

The contemporaneousness of these two disorders are medieval in acculturation, if not ancient. Mythopoeic and mystically preternatural, this belief displays a somewhat psychotic and bizarre delusional disposition.

The coetaneous theology, whether theosophical, Gnostic, syncretistic, Christian, Judaic, etc., possibly validated this diabolic condition. Zoanthropy is in modernity an accepted haecceity and indication of mental illness.

Synesthesia

In a physiological context synesthesia involves "sensation" experienced "in one part of the body while another part is stimulated." This coincides with sympathy neurologically in consideration of the nature of bodily parts to induce disorder or pain, etc., from one part and to have a similar effect on another.

In a psychological context this is a "process in which one type of stimulus produces a secondary, subjective sensation, as when a certain color evokes a specific smell" (1997 : 1358). LSD and other hallucinogenics can have this property -- the property of mixed sensations -- and synesthesia can simply be associating a form of sensual reference.

Hallucinosis

Concisely expressed, hallucinosis is "a mental disorder characterized by hallucinations" (1997 : 608). Auditory hallucinations are the most common. Hallucinations of vision, olfaction, gustation, and of touch or sensation are also experienced.

Sometimes distress, guilt, and/or substances can effect or precipitate this disorder. If brain damage is incurred I see no probable cure. If the causes are psychiatric, rather than organic, psychotropics have an efficacy to rectify this disquieting problem.

Echolalia

Consistent with the phonemic aspect of this expression, echolalia is an "automatic repetition of words," sounds, etc., by someone after the immediacy of hearing them.

This peculiar behavior is common in psychiatric hospitals and usually is "symptomatic of a mental illness" (Webster's, 1997 : 429). It is also common among the autistic.

I think a patient manifests this disorder to escape his present excogitations and in lieu of being somewhat imperiled by them. These reverberations offer relief, notwithstanding. Here a person is uncomfortable with his mental activity and tries to redirect it.

Metapsychology

Metapsychology is the speculation and study of the provenance, "function, and structure of the mind." It includes "the relationship between the mental processes and physical ones." This domain is "supplemental to psychology" and an important component of it" (1997 : 853).

Science is increasingly exploring the evolution, both neurologically and intangibly, of the properties of encephalic processes: how they affect body events, voluntary movements, involuntary movements, and what determines an individual's mental makeup. Why are no two minds the same? How do we delineate one mind from another?

Psychasthenia

Psychasthenic neuroses are an "old term for a group of neuroses indicated by phobias, obsessions, undue anxiety, etc." (1997 : 1085). These symptoms resemble paranoia albeit paranoid symptomatologies are concomitant with delusional and grandiose beliefs and are a psychotic phenomenon.

Psychasthenia can be treated with medication to alleviate its outward or positive manifestations. The remaining set of illness is apparently thought-related and requires therapy to minimize the gratuitous anxiety. This disorder can be stress-related or trauma-indicated, and is a serious illness.

Psychodrama

In "psychiatric circles [and] confines, psychodrama is a cathartic therapy" engaged by "one or more patients", who improvise or act out situations relevant to a personal or group problem. "These dramas are evaluated by therapists and fellow patients" (Webster's, 1997: 1085).

As an example, a patient may enact a former scenario such as a parent scolding this person needlessly. Ostensibly this individual has a troubling anemnesis of this event and a re-enactment will enable this party to deal with the memory emotionally and thereby expel some bottled-up psychic energy.

Ideomotor

This word is an interesting concept in "designating an unconscious bodily movement effected in response to an idea" (Webster's, 1997 : 670).

We are constantly enchanted by pleasant memories and disillusioned by unpleasant ones, and we act these emotions out in our body language. If we see a beautiful sunset, we are delighted. If we see a spider, we shudder. If I were to return from the bank and walk by a panhandler soliciting for money I might shrug my shoulders.

Hedonics

Hedonics are the "branch of psychology that deals with pleasant and unpleasant feelings" (1997 : 625).

The ancient Greeks founded and instituted various schools of philosophical and theological thought promoting eudaemonistic, ethical, and spiritual well-being. "Aristippus of Cyrene founded the Cyrenaic order of philosophy upholding immediate, sensual pleasure as the greatest good" (1997 : 345). The voluptuary Sybarites were "fond of self-indulgence and luxury (1997 : 1355).

Psychological pleasure is emotional pleasure. It includes happiness, contentment, and probity. Unpleasant feelings are unhealthy, painful, and an overt indication of conflict; they warrant consideration.

Negativism

In a psychological context, this condition is explicit as an attitude that ignores, resists, or opposes suggestions or orders directed from others (1997 : 907).

Whether friendly or authoritative, this contrary disposition leads to controversy, discord, and acrimony, on the job, in interpersonal relations, and is a tendentious manifestation of cynicism.

We as responsible cogs in life's machinery are disposed to frustrations and unfortunate pressures. When one component is faulty or vitiated, the whole sense of order and harmony is adversely affected.

Pick's Disease

"A. Pick was a Czech physician" who educated us in diagnosing, isolating, and defining an illness: Pick's Disease. It is characterized by "progressive deterioration of the brain with atrophy of the cerebral cortex, especially the frontal lobes, and is evidenced in loss of memory and emotional instability" (Webster's, 1997 : 1021).

The cerebrum, as a hemispherical process, is the two-lobed part of the brain, the largest brain portion, and it coordinates both memory and motor activities. It is the center of higher cognitive processes as well.

As the malady progresses the individual sometimes decays beyond a self-sufficing degree of autonomy and has to be institutionalized.

Facilitation

The psychological lexicon of this word concept can be defined loosely as progressive ease in any walk of life through which nerve resistance is lessened "by continued successive application of the necessary stimulus" (Webster's, 1997 : 485).

The stimulus, often a source of apprehension or inadequacy, is appropriately conducive to normal, healthy psychological functioning over time.

For example, on the job an individual may be fearful of performance (usually routine) with the employer proximal. Facilitation can also be conditioning and/or treatment in the case of some phobias.

Screen Memory

Part of a psychological mechanism in ego defense, screen memory is recognized in psychoanalytic circles as "a memory that can be tolerated and is used unconsciously as a screen against an allied memory that would be distressing if remembered" (Webster's, 1997 : 1206).

The use of memory as a function of a subterranean process of the mind prevents the recollection of prior events that lead to painful retrieval in remembering.

Often events are so painful that memory displaces them into the archives of the unconscious. The repression of one event to avoid another is also unconscious and precludes some dolorous anemnesis.

Cenesthesia

In psychology cenesthesia determines how we feel. It is the whole of undifferentiated sensations that reveal our awareness of the body and its condition, for example, the feeling of sadness, happiness, and illness (1997 : 227).

I presume these sensations are subject to mixed messages. One can be observed to experience pleasure and some sense of compunction, or feel these and other emotions.

These sensations are a manifestation of our emotional dispositions. Is feeling sad a physiological process as an emotional precursor or vice versa?

Sex Roles: Crossing Gender

It wasn't too long ago, as far as attitudes are concerned, that genders were consigned to specific roles. The man was dominant, emotionally aloof, and "either stolid or intellectual" (Slater, 1976 : 3), both at work and at home. By the same standards women were emotionally frail, unintimidating, and pragmatic in the home.

Men utilized their mate as an emotional outlet or valve, and women gained some societal standing in the background through man's social and economic status (Slater, 1976 : 4).

Through a trend in social awareness and a healthy emotional evolution in recognizing these obsolete parameters, we have attained a moderate degree of gender equality although many unhealthy stereotypes abound.

We no longer identify gender as a physiological consideration but an emotional, psychological one. This mode of modern gender attachment has led to confusion. An emotional man may be effusive although assigning a feminine gender may do conspicuous harm. We are nor prepared for this from a moral or social standpoint, although a unigender designation may prove itself preponderantly practical.

There are, of course, current attributes that are gender-distinguishable and one may avoid a sexist label although it may not be proper or display any sensitivity or tact in doing so.

Sexual Inhibition: No Longer

In a modern, sympathizing, and intellectual-oriented society there is a lessening of sexual inhibition: the sexual revolution, feminism, homosexual awareness, etc. We are, however, uncomfortable with "the banner of emotional expression" (Slater, 1976 : 5). Its message is instead of holding back feelings, we should be "up front" with them. This attitude, nonetheless, is daunting. And with those who are "sexually comfortable" with themselves, there are new hang-ups: "jealousy and guilt" (1976 : 5).

Jealousy, "a product of a capitalist upbringing" (1976 : 5-6), reveals a "possessive emotional bent" and even in the most extreme cases one is taught not to feel it.

And guilt has been arrogated as an inability to live in the present.

Younger Americans involved in relationships are constantly struggling to ignore moralities, such as "what Victorian women felt when sexually aroused", or what Quakers felt when angered. It may be said that people "feel guilty about feeling guilty," maybe for the first time (1976 : 6).

As long as feelings are suppressed, there is distress. And "the decay we attribute to 'old age' is the erosion of consigning humans to an emotional mold that does not fit" (1976 : 6).

Society: A Competitive Onslaught

Two currents of incompatibility and opposition in American Society are the self-serving "aggrandizement of the individual" (1976 : 9) in a competitive realm, and the need for people to live, love, and cooperate in a communal setting" (1976 : 8).

It isn't unconditional that Germans seek order, precision, and obedience and that Victorian Britain was stiff and Puritanical, but their societies are preoccupied with these values as a result of social conflict (1976 : 7). We as individuals should realize our diversity rather than suppressing these variations in order to paint a monolithic, uniform portrait (1976 : 8).

Some societal incompatibilities that warrant rectifying are "the invidiousness of our economic system" and "the extended family or local neighborhood." This dilemma which posits us seeking pleasure such as "symbolic virtues over our neighbors" and intimacy with them has made the appeal for communal "living more seductive" and the "need to suppress it more acute" (Slater, 1976 : 10).

The "flower child" of the 1960s was a burgeoning of these antipodean ideals and were not just demonstrating a rebellious mien. The flower children realized the futility, for example, of the war in Vietnam and the frustration of promoting a humanistic current that life is invaluable.

I will be brief here in deference to the sociological rather than the psychological current.

Alogia

Alogia is a negative (behavioral) deficit in schizophrenia and is evidenced by verbal blocking and lack of speech content (Davison & Neale, 1996, Glossary : 1). Alogia is symptomatic of catatonia notwithstanding the rigidity, the severity, and the rush of thoughts.

Often the patient is frightened, confused, and unaware of the environment, although these behaviors are a result of an emotional shock and a telltale sign of schizophrenia. Psychotropics are administered and these overt indications are remedied. The patient realizes his environment, is less frightened, and consequently feels safe and has a sense of community and belonging.

Blocking

A cerebral disturbance "often associated with thought disorders," this moderately problematic condition is characterized by an interruption in a train of speech by silence before an idea is fully communicated (1996, Glossary : 4). Blocking is not to be confused with some organic disorder in which thought processes are impaired, such as caducity, or brain damage incurred through injury or drug use, for example.

Blocking is simply an inability to maintain a course of ideation sometimes because of lethargy (lack of cerebral exercise) or a confusion of ideas, the selection of these ideas requiring effort or concentration.

Cohort Effects

These are the result and ramifications of living, experiencing, and learning in a given period with its particular "problems, challenges and opportunities" (Davison & Neale, 1996, Glossary: 5). Loosely, Zeitgeist. It is certain that the quality of life, the technological developments, the improvements, the progress, expedients of living in an advanced age are exceedingly favorable compared to ages heretofore. It is absolutely irrefutable, however, that for one reality, one advancement, that there must be preceding realities, etc. When we acknowledge the relationship and the relativism of a universal order, we can understand, appreciate, and value one another and our unique worth and importance.

Asociality

Asociality is a negative (behavioral) symptomatology in schizophrenic disorders, and is indicated by an inability to involve oneself in relationships and display any feelings of intimacy (1976, Glossary : 2).

Patients tend to lack self-esteem and consequently the confidence to maintain the social skills necessary in social settings, meeting people, and expressing affection to relatives, friends, and others.

Schizophrenics do isolate themselves and become self-contained, turned inward, and preoccupied with their own "private world." In these fantasies, having relationships would interfere with the afflicted person's special world.

Comorbidity

Comorbidity is the concomitance of two disorders, for example, drug-dependence and depression (1996, Glossary : 5). Many times these are related or they are mutually debilitating, or one problem develops into another.

Often if a disorder is manifest an individual will turn to alcohol, drug-use, gambling, or aggression, and abuse others especially if drugs or alcohol are being consumed.

Psychiatrists are able to treat most dual disorders. Many disorders are a manifestation of an underlying conflict in need of treatment. An individual with a nervosa is more than too large in their mind.

Choreiform

The word choreiform is cognate to "the involuntary, spasmodic, jerking motions" of the extremities and head that are "found in Huntington's chorea, and various nervous disorders" (1996, Glossary : 4).

Huntington's chorea is presenile dementia, inherited from a single dominant gene. Symptomatic are spasmodic jerking movements, psychotic behavior, and mental degeneration.

Parkinson's disease is characterized by uncontrollable muscular tremens, rigid gait, expressionlessness, and withdrawal.

Eugenics

Eugenics are the province concerned with ameliorating the heredity attributes or qualities of the human race to control breeding and reproduction (1996, Glossary : 9).

The Nazis of Germany and Moonraker in the James Bond film had this in mind when attempting to annihilate the planet in hopes of effecting a master race.

Humanely implemented, eugenics are currently utilized to engender an individual's high intellegence, athletic abilities, etc., without any genocidal intentions.

Epidemiology

This is the "study of the frequency and distribution of illness in a population" (1996, Glossary : 9). Frequency of schizophrenia in the United States is one percent.

Famine, and pestilence is more widespread in poorer countries.

AIDS is more frequent in urban settings in this society.

Alcoholism affects approximately 90 to 100 million people in this country.

Alter Ego

According to DSM-IV a precise diagnosis of Dissociative Identity Disorder requires that the individual "have at least two ego states or alters" (Davison & Neale, 1996: 180).

These alters are defined with their particular behaviors, memories, and relationships. These personalities as different entities, facets, may indicate "different handedness, wear glasses of different prescriptions, and have different allergies" (1996 : 180).

The frequency of DID has spiraled upward ever since "The Three Faces of Eve" and "Sybil" were published.

Body Dismorphic Disorder

This disorder is defined by a preoccupation with an individual's "imagined or exaggerated defect in their appearance" (1996, Glossary : 4).

Sometimes a patient is obese and arrogates this as the real issue, problem.

Other times, as in anorexia nervosa, a patient is erroneous in his or her perception of the self.

BDD often develops into a full-blown obsession.

Clonic Phase

In a grand mal epileptic attack these are the occurrence of violent contortions and jerking movement (1996, Glossary : 5). Although epileptics are given medications it is important that those afflicted are afforded a safe environment in guaranteeing the least possible event of harm for all considered.

Essential Hypertension

As a psychophysiological disorder, essential hypertension has no organic course. Over time this develops into "enlargement and degeneration of small arteries, kidney damage, etc." (1996, Glossary : 9).

Type A personalities I would suspect may invariably incur this malady.

Anxiety and anger may also contribute to it.

Stress appears to be a greater factor.

Sexual Orientation Disturbance

Another and "earlier term for DSM III's ego-dystonic homosexuality" (1996, Glossary : 23), this condition is a result of social and cultural consideration, attitudes as well as ego-personal ones for homosexuals.

Homosexuality is an aspect of almost all societies. It is accepted in many; some accept homosexuals but not homosexuality; and some cultures denounce this practice. Some claim homosexuality is a choice and some homosexuals unequivocally maintain some hormonal contingencies.

Response Set

This disorder is individuated as the tendency of an individual to respond in a certain way to questions or statements on an exam. For example, the respondent may answer false notwithstanding the content of validity of each question (1996, Glossary : 22). This apathetic mode of involvement demonstrates a remoteness, disregarded for priorities (what is important) and detachment. It is indicative of a major disorder or depression.

Response Acquiescence

Instead of the negativity in responding to questions, a person with response acquiescence agrees with any questions regardless of content, as a yes-saying response set (1996, Glossary : 22).
This demonstrates practically the same apathy, detachment, and heedlessness.
Goals are what keep us directed, and give us meaning. When we lose direction, our lives lose their meaning.

Right to Refuse Treatment

As far as legal parameters are concerned, a committed patient has the option and the right to refuse risky or unconventional treatment (1996, Glossary : 22).
This is a true example of medical ethics in a free society.
Many psychiatrists want to assume more authority in similar cases of medicine vs. legal rights. Many patients experience unduly the pain, suffering, and trauma characteristic of psychotic, depressed states although they have that right. Ask Kevorkian.

Self Psychology

This therapy is "Kohut's variation of psychoanalysis." The focus here is on the individual's sense of self-worth through approval and support by "key figures during childhood" (1996, Glossary : 23).

Early childhood is crucial phase in development. With the child's parents, siblings offering love, guidance, nurturance and a stable environment, a healthy, mature, and responsible individual will emerge.

In looking back we remember dominant, important figures, people we admired and loved. Childhood happiness will usually determine whether or not a person will need this therapy.

Systematic Desensitization

"A major behavior therapy procedure that has a fearful person, while deeply relaxed, imagine a series of progressively more fearsome situations. The two responses of relaxation and fear are incompatible and fear is dispelled. This technique is useful for treating psychological problems in which anxiety is the principal difficulty" (1996, Glossary : 25). Another therapy for anxiety or phobias are the closer proximity encounter method whereby successively closer proximities obviate the otiose and debilitating fears.

Sociogenic Hypothesis

This doctrine upholds the concept that having low socio-economic status causes schizophrenia (1996, Glossary : 24).

I believe having poor economic standing may contribute to a higher incidence of disease because of medical costs. I do not believe, however, that economic conditions affect the frequency of schizophrenia. I do think that poor people do not have the means to rectify or treat the illness with that self-same capacity of higher income families.

Wealthier individuals, if mentally ill, are provided with institutions that are monetarily more exclusive and therefore more opulent. There is no price on illness.

Tarantism

The tarantism is a frenzied dance mania, widespread in the 1200s, purportedly generated by the bite of a tarantula (1996, Glossary : 25).

The venom of a tarantula usually is not deadly, particularly in adults, although any poison is harmful.

There is a similar dance of that or of a different time frame. This dance shares a similitude with that name, and this wildly frenetic movement is designed to egest the poison. The music to the dance is a composition in arachnoid rhythmicity.

Vicarious Conditioning

Vicarious conditioning is learning by watching how others react to stimuli or by listening to what others say (Davison and Neale, 1996, Glossary : 27).

Somewhere in time and space there is an apothegm: learning is observing or vice versa. This campaign into the relational continuum assures that individuals share, cogitate, and interact with social consistencies and parameters and that societal mores are maintained.

Voodoo Death

The death of a member of some tribe or uncultivated peoples after violating some tribal law or being imprecated by a witch doctor (1996, Glossary : 27).

The implementation and effectiveness of this much feared and credulous punishment, according to a psychologist I had engaged (no betrothal here), is in the daunting faith in which voodoo is considered real.

How powerful is idol worship?

Advanced Accurate Empathy

This ilk of empathy is "one in which the therapist infers vital concerns and emotions, feeling" that are essential in what the client is discussing. It is an interpretation (1996, Glossary : 1).

In therapy it is important that the professional can understand what specific feelings are involved as an emotional bridge in a client-therapist relationship.

Because emotions are powerful and revealing, their recognition is important in effectively treating their unhealthiness and indications.

Narcosynthesis

This psychiatric procedure was first implemented during WWII. It constituted the administration of a drug that enabled "stressed soldiers to depict battle traumata" (1996, Glossary : 16) in order to ameliorate disordered conditions.

With any war, lives, whether civilian or military, are compromised and inordinate stresses are incurred. Many times these leave a legacy of painful, frightening, and utterly distressing psychological problems.

Was it chance that I first exhibited psychotic tendencies after signing up for the draft?

Primary Empathy

Another form of empathy, primary empathy occurs when the therapist comprehends the gist and feeling of what the patient is alluding to and what is "expressed from the client's phenomenological point of view" (1996, Glossary : 19). The phenomenology is what particulars the client experiences.

This empathy enables the professional to have explicit knowledge of what feelings and how these feelings are affecting the emotional bent, disposition of the client. Thus greater understanding is communicated and treatment is accessible.

Morbidity Risk

Morbidity risk is the likelihood that a person will develop a specific disorder (196, Glossary : 16). With the recent barrage of genetic disclosures in determining illness, longevity, and manifold dispositions, it is extremely plausible that genetics are responsible for our psychophysiological destinies.

There are other considerations as well such as environment, stress, and contact, etc.

First-Rank Symptoms

In defining schizophrenia, there are positive symptoms, such as specific delusions and hallucinations. As put forth by Schneider, these are significant for a more exact diagnosis (1996, Glossary : 10).

When an individual is psychotic, overt symptomatologies such as persecutory ideations, bizarre beliefs like controlling global weather patterns or having an electric but non-material mind, and hearing in the next room Adolph Hitler being heralded by a frantic and vociferous crowd, seem as real as possible although the affected individual may realize the psychosis in the back of his mind.

Grandiose Delusions

This symptomatic existent sustained in paranoid schizophrenia and other forms of paranoid disorders, is "an exaggerated sense of an individual's importance, power, knowledge, or identity" (1996, Glossary : 11).

Often the patient deems him or herself the most attractive individual, is being sought by the populace, is psychokinetic, is the reincarnation of someone famous, or has some exclusive knowledge and is being tracked by the CIA.

Genuineness

This is a client-oriented therapy and is an essential feature of the "therapist that refers to candidness and authenticity" (1996, Glossary : 11).

An empathic and concerned therapist can gain the trust and fashion a setting conducive to ministration. The client is consoled and is able to express many profound feelings. This ambient sense of trust is necessary, therapeutic, and fundamental.

Flight of Ideas

These are symptoms of mania distinguished by a quickened shift in conversation from one matter of topic to another with only cursory associative connections (1996, Glossary : 10)

In a manic episode thoughts are usually spontaneous. The mind is in high gear and there is no time for an intercession of deliberation, judgment, and or ratiocination. The manic person may spout off about aliens running the government and then shift automatically to talking about radio waves causing mutations.

Expressed Emotion

This phenomenon is the substance "of criticism and hostility" attributed by others to the patient "in the literature of schizophrenia, especially within a family" (1996, Glossary : 10).

Concomitant with the illness are behaviors (negative symptoms) that are considered bizarre, for example, isolation, unresponsiveness, strange movements, proneness to anger, laughing or crying. With these symptomatologies it is imperative that the patient be committed to the care of professionals until he retains a relatively guarded degree of stability.

Confabulate

In psychology this is the process of introducing in gaps of the memory detailed versions of unreal events believed genuine by the narrator (1996, Glossary : 5).

Oftentimes the mentally ill instead of recognizing a hiatus in their past memories will contrive, for example, playing parts in movies, being married to a lawyer, etc. Some patients are inordinately absorbed in fantasy. Some patients do not remember their own name.

This malady is mostly an anomaly of the older set. Sometimes fantasy or pretensions are more tolerable and esteemed psychologically.

Aphasia

This devitalizing malady is "the loss or impairment" of the capacity to utilize language incurred by "lesions in the brain." The executive designation in effect is characterized by difficulties in verbalizing or inscribing the words intended. The receptive component is characterized by difficulties in comprehending inscribed or nuncupative language (1996, Glossary : 2).

There can be some degree of functioning independently although the damage may require that the impaired be institutionalized.

Cognitive Restructuring

Cognitive Restructuring is any behavior therapy measure that undertakes to modify the mode in which a client views life in order to rectify overt behavior and detrimental emotions (1996, Glossary : 5).

Cognitive Therapy (CT)

"A cognitive restructuring therapy associated with the psychiatrist Aaron . Beck," CT is "concerned with changing negative schemata (mental structures for organizing information about the world) and certain cognitive biases or distortions that influence a person to construe life in a depressing way" (1996, Glossary; 5).

Beck upheld that depressed individuals are that way because their ideations are biased toward negative interpretations (Beck's Paradigm). The following are levels of cognitive processes that Beck attributed to depression (Davison & Neale, 1996 : 231).

- "Negative Triad (Pessimistic Views of Self, World and Future)
- Negative Schemata or Beliefs: Triggered by Negative Life Events (e.g., the assumption that I have to be perfect)
- Cognitive Biases (e.g., arbitrary inference)
- Depression"

Beck believed we acquired a negative schema while enduring, for example, the loss of a loved one, or being rejected by peers. These beliefs or schemata are rekindled by situations that resemble, sometimes remotely, the ones that shaped the original bias. The learned negative schemata "fuels certain cognitive biases" that make suffers misapprehend reality.

These unhealthy biases, along with cognitive distortions encompass "Beck's negative triad, negative feelings of the self, the world, and the future." Following is a list of the primary "cognitive biases of the depressive person."

(1) "Arbitrary Inference" (Davison and Neale, 1996; 231). A belief not grounded on sufficient evidence; for example, an individual thinks he is to blame for it raining on his day to go fishing.
(2) "Selective Abstraction." A conclusion selected from one possibility of many different interpretations, such as when a worker feels responsible for a defective product although he or she is but one who contributed to its manufacture.
(3) "Overgeneralization." An encompassing judgment inferred from a single case, for example, a student feels worthless and stupid because he does poorly on one test.
(4) "Magnification and Minimization." Exaggerations in rating performance. For example, a man when painting his house gets some paint on the window and feels he has ruined his home (magnification) or a woman who wins a prize for baking cookies believes that she didn't even come close to deserving it (minimization) (1996 : 233).

Somatoform Disorders

These are disorders characterized by physiological symptomatologies and point toward a physical problem but they are manifested in light of no physiological etiology. It is accepted, through not undeniably, that they are fashioned by psychological conflicts and needs although not a voluntary process in individuals (1996, Glossary : 24)

Much of the etiology of somatoform disorders were aimed at Freud's concept of hysteria. This has encouraged ideology that both conversion disorder and somatization disorder are alike (Davison & Neale, 1996 : 171). More on conversion and somatization disorder will follow.

Three categories of somatoform disorders in DSM-IV include:

Pain Disorder. Here the patient incurs extreme and prolonged pain that has no association or linkage to organic pathology. It inclines to be stress-related or enables the individual to avoid antipathetic activities. Sometimes it is effected to obtain sympathy and attention (1996 : Glossary; 18).

Body dysmorphic disorder: See above.

Hypochondriasis: "A somatoform disorder in which the person, misinterpreting rather ordinary physical sensations, is preoccupied with fears of having a serious disease and is not dissuaded by medical opinion. It is difficult to distinguish from somatization disorder" (1996, Glossary : 12).

I think the hypochondriac is morbidly preoccupied with fearing the worst possible contingency: death.

Conversion Disorders

These are somataform disorders in which bodily functioning is impaired to a significant degree, indicating a neurological illness. Although there is no bodily, organic damage, disease, sensory or muscular functioning indicates otherwise (1996, Glossary : 6).

Symptoms include "paralyses of arms or limbs; seizures; coordination disturbances; prickling, tingling, creeping on the skin; insensitivity to pain; loss or impairment of sensations called anesthesias" (Davison & Neale, 1996 : 167).

Conversion symptomatologies indicate they are psychological and appear under stressful conditions. Freud derived the term from "the energy of a repressed instinct to be diverted into sensory-motor channels and block functioning." This emotional and psychological conflict is thus "converted" to physiological symptomatologies (1996 : 167).

Some conversion disorders:

Anesthesias: "an impairment or loss of sensation usually of touch but sometimes of the other senses." Symptoms include impaired vision; partial or complete blindness, and/or tunnel vision (Davison & Neale, 1996, Glossary : 2).

Aphonia: loss of voice except the ability to whisper (Davison & Neale, 1996, Glossary : 167).

Anosmia: "loss or impairment of the sense of smell" (Davison & Neale, 1996, Glossary : 2).

False Pregnancy

Hysteria: "A disorder known to the ancient Greeks in which a physical incapacity, a paralysis, an anesthesia, or an analgesia, is not due to physiological dysfunction. For example, glove anesthesia, an older term for a conversion disorder. In the late nineteenth century dissociative disorders were identified as such and considered hysterical states" (1996, Glossary : 12). More on dissociative disorders will follow.

Analgesia: "an insensitivity to pain without loss of consciousness, sometimes found in conversion disorder (1996, Glossary : 2).

Paresthesia: "a sensation of tingling or creeping on the skin" (1996, Glossary : 18).

Somatization Disorder (Briquet's Syndrome)

Definitely speaking this is "a somatoform disorder in which the patient continually seeks medical help for recurring and multiple physical symptomatologies that have no discoverable physiological cause. The medical history is complicated and dramatically presented" (1996, Glossary : 24).

Factitious Disorders

As a DSM category these are disorders manifested "in psychological or physiological symptomatologies that seem under voluntary control and are assumed by the individual" to appear sick. This desire to appear sick is not in fact voluntary, and is an indication of a severe illness (1996, Glossary : 10).

This disturbance is severe in that its motivations are not clear. It is possible that the individual wants to be a patient, and these desires fashion the disorders. These are classified conversion disorders (Davison and Neale, 1996; 169).

Malingering

Malingering is feigning "a physical or psychological illness in order to avoid responsibility" or because of some other motive. Unlike a factitious disorder, the goal here is recognizable (1996, Glossary : 14).

This disorder is diagnosed when conversion-like symptoms are deemed unequivocally voluntary on the part of the patient (1996 : 169).

Discriminating malingering from conversion disorders is difficult if not impossible because clinicians are not clearly certain whether "behaviors are consciously or unconsciously motivated" (1996 :169).

La Belle Indifference

This is a behavior that enables the clinician to distinguish between malingering and conversion reactions. It is characterized by an attitude of unconcern on the part of the affected person. This unconcern is incompatible with the severity of the symptoms and "long-term consequences."

"One-third of conversion disorder cases demonstrate la belle indifference" (1996 : 169).

Dissociative Disorders

Dissociative disorders are marked by an individuals changing sense of "identity, memory, consciousness." Those affected may forget "important individual events," "their identity," or take upon themselves a new identity (1996 : 178).

"Prevalences of 7.0 percent, 2.4 and 0.2 were determined for amnesia, depersonalization, and fugue, respectively, according to Ross in 1991" (1996 : 179).

Four dissociative disorders -- dissociative amnesia, dissociative fugue, dissociative identity disorder, and depersonalization disorder are discussed here.

Dissociative Amnesia

The individual is unable to remember significant personal events. This is characterized, for example, by memory loss for all occurrences in a limited time period that follows some trauma.

An etiological example is the death of a loved one. Less common is amnesia for selected events of a "circumscribed period of distress" (1996; 179).

With total amnesia, the affected person does not recall relatives and friends, but remembers how "to talk, read, reason, may retain talents," knowledge of the world, and how to function. "The episode might last several hours or as long as years" (1996 : 179).

Dissociative Fugue

Previously examined. Additional commentary.

The fugue not only forgets (becomes amnesic) but "leaves home and work and acquires a new identity." He or she may assume "a new name, new home, new job," and new personality. He or she may establish a "complex social life."

The duration of the "new person" disorder is comprised of parochial but "purposeful travel" and few contacts.

Trauma (severe stress) such as "marital quarrels, rejections, military service, or a natural calamity" may provoke an episode. Recovery in varying time periods is complete without the individual recollecting the flight and what precipitated it (1996 : 179).

Dissociative Identity Disorder (Multiple Personality Disorder)

Previously examined. Additional commentary.

DID usually has its roots in childhood although a few cases are diagnosed before adolescence. DID is considered to "be more chronic and severe than other dissociative disorders."

It occurs in women more so than men. Recovery may be less complete than other DIDs.

DID is often accompanied with other diagnoses, for example, "depression, borderline personality disorder, and somatization disorders."

"Headaches, substance abuse, phobias, suicidal ideations, and self-abusive behavior" is also evidenced. DIDs are two or more "separated and coherent systems of being." Schizophrenia is a "split between cognition and affect" (Davison and Neale, 1996 : 180).

Depersonalization Disorder

Although controversial as having its inclusion in DSM-IV as a dissociative disorder, depersonalization disorder is the "perspective or experience" of the self as a bizarre and disruptive alteration.

During an episode the individual loses his sense of reality of the self. The body becomes distorted or they may feel they are outside of themselves (1996 : 182). During a personal episode I thought/perceived giant growths all over my body. An individual also may feel everyone is a robot or feel that he is in a dream.

A schizophrenic, by contrast, does not demonstrate the "as if" quality of the disorder. A schizophrenic feels "real and complete" estrangement from his being (1996 : 182).

Etiology of Dissociative Disorders

Learning theorists share the view that dissociative disorders are manifested as "avoidance responses." These behaviors "protect the individual from highly stressful events." Here the behavioral view and psychoanalytic orientations are similar.

Bliss (1980) has determined DDs to be "established in childhood by self-hypnosis" in order to "cope with extremely disturbing events." High rates of "physical abuse in childhood (80 percent) and incest (70 percent) were indicated in a survey by therapists" (1996 : 182-83).

Sexual Dysfunctions

These dysfunctions are delineated by inhibitions in the normal sexual cycle through appetitive or psychophysiological changes in the individual.

Sometimes our most memorable experience of the sexual act is during youth and somehow over the years sex loses its thrill, its feeling of insuperable yet momentary physical and emotional pleasure

I think in men and women the foreskin gradually loses responsiveness, excitability, our hormones change, and/ or we just run out of libido over time. Nonetheless Mae West was active in her eighties.

Selective Mortality

"A possible confound in longitudinal studies, whereby the less healthy people in a sample are more likely to drop out over time" (Davison and Neale, 1996, Glossary : 23).

Longitudinal Studies

These are studies that reap information from like candidates intermittently, sometimes for years. These experiments are employed in order to determine how patients' realities change (1996, Glossary: 14).

Confounds

These are contingencies or variables that are so variegated and intermixed that no accurate mensurability exists separately. Confounds thwart the data retrieved as imprecise, and make unequivocal determination impossible (1996, Glossary : 6).

Psychology, abnormal psychology, and many other sciences are not implicitly/ explicitly ideal or perfect. Information is extrapolated and gleaned from antecedent cases, human observations, generalities, biases, common sense, state-of-the-art design and experimentation techniques, and the ever-increasing onslaught of defining, understanding new disorders and illnesses. I think so far the field has prospered and will continue to improve.

Paraphrenia

This term is recurrently applied to cases of schizophrenia in older individuals (1996, Glossary : 18).

The usual onset of schizophrenia is in young adults during the transition between puberty and adulthood.

Once afflicted, a schizophrenic is liable to remain a schizophrenic and require treatment throughout the course of his or her life.

A paraphrenic can also be treated although successful chance of recovery is less likely, especially in consideration of negative symptomatologies (flat affect, apathy).

Negativism Reexamined

Tendentiously arrogated, this behavior is characterized by acting in opposition to the desires and expectations of others (G-16).

This behavior is indicated by the individual's cognizance. It is a form of passive-aggression, and clearly has its foundations in asocial comportment and demeanor. An overt example is using abusive language in church or more subtly, laughing at the death of an assassinated world leader.

Occipital Lobe

"The posterior area of each cerebral hemisphere, situated behind the parietal lobe and above the temporal lobes, responsible for reception and analysis of visual information and for some visual memory" (1996, Glossary : 17).

Occasionally we observe an individual who has used a gun in a suicidal attempt. Unfortunately such attempts result in irreversible damage incurred without killing the shooter.

Most gunshots to the brain in a suicidal attempt are proximal to the frontal and temporal lobes. Abraham Lincoln was shot in the occipital area.

Proactive

This term means "relating to or caused by previously learned behavior, habits, etc_ (proactive inhibition)" (Webster's, 1997 : 1072).

Quite frequently a patient is hospitalized for mental illness, and is tested for organic or physiological causes and origins. The patients are given ECTs, checked for hearing disorders, and tested by a neurologist. Inevitably their disorder is deemed a psychological phenomenon and thus requires medications and/or behavioral therapies.

Mourning Work

Freud used this terminology in his theory of depression. It is defined as the recollection by a depressed individual of memories of a deceased one. This pragmatically, necessitously serves to detach or separate the person from the deceased (Davison & Neale, 1996, Glossary : 16).

In moments of memory and solitude we reflect reminisce. We have all been bereft through losses of friends and family. In realizing the ineluctable eventually we are compromised. These memories, however, remain and render us human, emoting and rational beings.

Orgasmic Variable

These are the agencies, determinates of a physiological or psychological context that are deemed to be functioning "under the skin." These are utilized as means of behavioral assessment (1996, Glossary : 17).

In assessing a diagnosis it is useful to understand the motives and desires at a more unconscious level.

As a means to rectify behavior, it is both practical and requisite for the therapist to be aware of why the patient is angry, fearful or depressed. Often there are a myriad of factors that merit probing in order to understand a behavior.

Nervous Breakdown

Touted as a psychotic or neurotic disorder that interferes with or impairs an individual's normal functioning (Webster's, 1997 : 910), a nervous breakdown is a catch word, a euphemism for an episode that requires hospitalization.

Dyer (Your Erroneous Zones, 1996) describes it concisely: "Your nerves don't break down." Often is the case, however, is that we as consumers or employees are compelled through societal standards to work hard, to over achieve. Sometimes we just collapse emotionally, the mind and body needing a respite from these enervating and deleterious stressors called life.

Myxedema

Myxedema is characterized by "a thyroid deficiency in which proper metabolism is slowed down" resulting in the patient becoming lethargic, thinking more slowly and experiencing depression (Davison & Neale, 1996, Glossary : 16). Also described as hypothyroidism, the patient, because of obesity is depressed and is obese because of depression. Therapy, medications and hospitalization is necessary.

Loose Association (Derailment)

A facet of thought disorder in schizophrenia, derailment is defined as a breaking off of the main train of thinking usually because of difficulty in staying with the same topic or theme of conversation, and also due to memories of the past (1996, Glossary : 14). Loosening of associations is a lack of coherent and unwavering focus on a subject. It is a telltale indication of schizophrenia (negative symptomatologies). Schizophrenics, because of the very nature of the illness, have a hard time carrying on a conversation without a social blunder of some sort.

Lithium Carbonate

Lithium is a medication prescribed for mania and depression in bipolar disorder, also known as manic-depression (1996, Glossary : 14). During the manic phase there is intense excitement, euphoria, and poor judgment exhibited.

The depressed phase is indicated by, for example, psychic pain, apathy, hopelessness, and suicidal tendencies. In treating the illness the goal is to balance the highs and lows and provide the patient with some psychological and emotional median. The manic aspect can be treated and rectified in a relatively short time. The depressed phase sometimes takes months to regulate.

Medical (Disease) Model

This is an application in abnormal psychology that conceptualizes abnormal behavior in a similar light to physical diseases (1996, Glossary : 15).

Abnormal behavior is manifested appropriately from the source of an abnormal psyche (chemical imbalance). What causes these dysfunctional psyches is what the field of psychiatry attempts to explain, reveal and remedy. Abnormal behavior is categorized and organized for the sake of scientific mindfulness, assessment, definitive reductionism, and remedial contingencies.

Midbrain

"The middle part of the brain is comprised of a mass of nerve fiber tracts connecting the spinal cord and pons, medulla, and cerebellum to the cerebral cortex" (Davison & Neale, 1996, Glossary : 15).

The brain as a medical scientific and organic entity and its processes receives a greater degree of attention than any other field of human concern.

As an integrated, organized, and profoundly complex whole, it surpasses any sate-of-the-art computer in its sophistication.

Phonological Disorder

This malady is indicative of a "learning disability" wherein the disordered person pronounces words sounding like baby talk because he or she is unable "to make certain speech sounds"(1996, Glossary : 19).

Often peers deride or tease the affected person, reinforcing notions that he is different and thus concomitant psychological conflict becomes manifest in the individual; one feels excluded, inferior. We as adaptive, survival-oriented organisms have an unfortunate way of living up to our expectations of others. I think with phonological disorder, the person is deemed unintelligent, his other abilities notwithstanding. To accept oneself is the crux of the solution.

Prevention

"Primary efforts in community psychology to reduce the incidence of new cases of psychological disorders by such means as altering stressful living conditions and genetic counseling; secondary efforts to detect disorders early, so they will not develop into full-blown, perhaps chronic, disabilities; tertiary efforts to reduce the long-term consequences of having a disorder, equivalent in most respects to therapy" (1996, Glossary : 19).

With holistic medicines and technologies in their heyday, preventive measures seem the best means to combat illnesses.

Addison's Disease

This is an endocrine-related illness that is characterized by a cortisone inadequacy. Loss of weight, fatigue and melanism are outward manifestations (1996, Glossary : 1).

A.D. shares symptoms with other illnesses. Weight loss and hebetude can be a psychological consideration or a physiological illness; a sign of infection or virus. Melanism is common in liver problems or aging, or from medications and other causes.

Woolly Mammoth

This terminology refers to the unconsciousness of the individual in which hidden "conflicts of psychoanalytic theory are encapsulated." This makes the conflicts unavailable in analysis and adaptation; if not remedied, these conflicts develop into serious disorders in adulthood. (Davison & Neale, 1996, Glossary : 27). If psychic conflicts are a source of behavioral and emotional trouble in younger persons, it is imperative that through hypnosis or some sort of suggestion these tormenting dissonant processes are discovered and treated.

Changing behaviors manifested from these conflicts in older populations is exceedingly more difficult to rectify.

Working Through

In psychoanalysis, this is the laborious, lengthy procedure by which the analysand deals with repressed conflicts time and time again and owns up to the analyst's interpretations until these unhealthy issues are sufficiently treated (1996, Glossary : 27).

Again crucial to the science and efficacy of psychoanalysis is the professional's persistent and circumspect disclosure of the source of the unwholesome unconscious processes. If the focus on the conflictual and subterranean processes is clear, they can be brought to the surface so they can be compromised and dealt with adequately.

Transvestitic Fetishism

This is the behavior of wearing clothes of the opposite sex. It commonly evokes some degree of sexual arousal (1996, Glossary : 26).

Often we hear about and take notice of transvestites in city settings. Some may work at clubs entertaining or engaging in prostitution. Despite its strange connotations many normal men actually enjoy gender-crossing and this phenomenon has gained some public acceptance. More acceptable are women who share this condition.

Although this form of fetishism indicates homosexuality for both genders, it is not an absolute certainty.

Thanatos

In Greek mythology Thanatos is death personified; the Roman Mors is recognized thus as well (Webster's, 1997 : 1385). As a function of the id this concept is the death instinct. Eros is the other basis instinct (Davison & Neale, 1996, Glossary : 26).

Freud and other innovators considered this death wish as fundamental to our being. There are sexual implications contingent and the final act in one's existence can be a fantasy fulfilled in gratifying some sexual urge or erotic impulse.

Social Selection Theory

This interesting theory attempts to explicate the relationship between the social status and schizophrenia in that schizophrenics are lowered in societal standings (1996, Glossary : 24).

We are all aware there is a stigma associated with mental illness, as there is in accord with some visible physical disability.

Many individuals are daunted, or experience panic when faced with the prospect of visiting a mental institution.

The red mark on one's forehead is somewhat diminished as we acquire insight and acceptance in a society that takes care of its disabled and afflicted.

Systems Perspective

This is a general viewpoint or belief that considers, for example, that a youngster's behavior problems should be examined in the context in which they occur such as in the environment of the family unit and the child's school (1996, Glossary : 25).

This is important as there are many variables that require being explored and analyzed in drawing a profile of the affected individual. It isn't so much an inherent problem or phenomenon, but more an interpersonal, social issue, especially in the child's world.

If untreated, the child may develop as emotionally troubled.

Subintentionable Death

This phenomenon is defined as a death that has no conscious motive but may be the result of some unconscious process or intention (1996 Glossary : 25).

An example is reckless driving by an individual who just had an argument with his girlfriend. Overtly, consciously, he is unaffected although the shock may have a delayed effect or be submerged into the psyche.

Many times the prospect of dying is so frightful to the individual that there is no awareness of the desire to die.

Orthopsychiatry

"This is the study and treatment of disorders of behavior and personality, with emphasis on prevention through a clinical approach" (Webster's, 1997 : 957).

Unlike various schools of psychiatric science, theory and their attendant methodologies in rectifying unhealthy symptomatologies (Gestalt, psychoanalysis, neuraleptics), this mode of treatment through the use of previous case observation, studies, and experience in the field can in effect alter or check the course of the disorder. For example, if an incipient illness or disorder is recognized and diagnosed, the proper therapies and medications can lessen and mitigate the debilitating and daunting effects. A manic episode and psychosis can thus be dealt with preclusively.

Exhibitionism

"A designation in psychology, exhibitionism is a tendency to expose parts of the body that are conventionally concealed, especially in seeking sexual stimulation or gratification (Webster's, 1997 : 476).

Exhibitionism is a perversion, legally and medically defined and adduced; the "flasher" longs for attention, expressly female recognition. This inappropriate activity involves the object (female) as the "sexual nexus" and often has a traumatizing effect on the object of the act.

Streaking, the fad of the 1970's on the other hand, differs to the extent that those observing the streaker have a choice. Nonetheless, exhibitionism and streaking are illicit activities.

Monomania

A mental disorder, monomania is distinguished by an irrational preoccupation with one subject (Webster's, 1997: 878). This preoccupation interferes with life's demands of employment, study and interpersonal relationships. Often a monomaniac can function in a manifold of usual practices albeit the individual is not able to grow, develop emotionally and psychologically.

A preoccupation with acquiring money or status is common but an inordinate preoccupation with that is unhealthy and problematic.

Folie de Grandeur

Simply defined this French expression means "delusions of grandeur; megalomania" (Webster's, 1997 : 523).

Delusions of grandeur properly defined are beliefs that have no basis in reality in which and individual trusts he or she is to achieve or has a prodigious, fantastic, divine wealth of one or more attributes.

A megalomaniac believes he has power, wealth, and manifests delusions of grandeur. These manifestations are indications of a distorted, even psychotic state and can be treated with medications. Although not a criterion for involuntary confinement in a state-run facility, many patients in them and in private hospitals exhibit folie de grandeur.

These two short stories are indicative of psychotic episodes and were written after living through personal experiences.

WALT IN HIS ORBIT

The room was a light turquoise as he meditated at twilight. It was small, what he considered to be the living room. Across from him was the couch where he kept his weights. He had just finished reading and placed the books on the stand next to his medication. He was a lonely youth and spent much of his time musing, fantasizing about being handsome and cool, with women looking at him in admiration. Walt lived in a poor section of town.

"Enough," he commanded himself and quickly looked up at the only picture on the wall. It was a sketch, a rustic scene of an amorphous figure at a mailbox during winter. In color, it was a blend of brown, fuschia and azure with warm conservative tones. It was inviting and eternal, sort of a lost spatial network of formlessness in the realm of a dwelling, this room, where the infinite is separated by an enclosure, the individual as the ultimate discernor of what is real and what is imagined. Drifting once more, he thought about time, forces in the past, and those in his predicament, longing for some equation. Those in his situation long ago were removed, locked up, often chained, and poorly fed.

Soon he was absorbed in reverie, mulling about the future, meeting a girl who was at his level in value, and acquiring a car. Coming back to tangible reflection, he opened up his stand and removed some marijuana. After putting a meager amount in his pipe, he lit up the bowl and inhaled. Walt got high several times a week although it tended to make him wary of others.

After a few coughs, our fellow began feeling its effect, that numb gratification, and started to unwind.

It was after that certain puff, just as the drug took effect, that Walt would sense a small amount of suspicion and ponder grand thoughts although they were profoundly unreal.

Delivering himself from bewilderment he gazed through the yellowed curtain noting a premonitory darkening of the horizon. His rumination was soon replaced by apprehension.

A sudden streak of light was followed instantaneously by a booming clap as a torrent of droplets pounded the street below. Walt imagined the water rinsing the pavement of grease, papers and dirt, a flowing stream of residue being filtered tirelessly through the sewer grates, that abject area of communal disregard.

Another flash and our alarmed hero stirred in fright but kept most of is composure realizing the storm's insubstantial effects on his sheltered existence.

A faint knock interrupted Walt's immediate involvement. Trying to assemble the events, he realized there were no footfalls and dismissed the noise as illusion.

Walt Crimmons didn't perpetually believe himself as a loser. At one time he was enrolled at a university and to a certain degree well-received.

The greater part of college females, Walter presupposed, were pristine and outnumbered the male commonality especially when it came to chastity. But having desire to pursue gratification through drug use undermined what industry he employed in study or relationship.

The drug Walt took made him fall asleep and he woke to find himself supine on his couch with the radio blaring. "Someone else inhabited my apartment. I don't think I sleepwalk," he reasoned. He hoped it was a girl who made the arrangement but this was doubtful.

All Walt could do is run and get away from the bizarre occurrence, but why would someone be tracking me, he thought. He was an innocuous character and no one of stature. "What if they meant me harm?" he asked himself; but he couldn't compromise the situation not knowing the twisted motivation as he kept his sanity over the predicament.

Maybe he was used as some strategic plan by the police or F.B.I. They start on the little guy so maybe he can confess to nail the biggies.

Logic interceded and Walt concluded that someone wanted to scare him. "But why would they trouble themselves with me?" he thought.

He remembered. It struck a painful chord. He knew why.

LOST IN L.A. 1993

The driver's face melted as he turned to say "Be a man," but was only an illusion. Gary became tearful and the Mexican passengers seemed to sympathize as he got off the bus. He wanted to walk the rest of the way if only to feel safe, especially after that incident with the lives he saved, the bus bouncing those people around on that sharp left, a few days ago. -- He overestimated his strength.

Now this afternoon, Gary felt brave enough to turn left on Hollywood and walk a fair distance along Robertson. His destination was one set of apartments where he believed he was accepted. "Buy some herb," one of the denizens suggested as he spotted. They spent the entire afternoon smoking his expense.

Night crept up on their diurnal activity and our man started his journey home. There it was in scarlet letters, glaringly obvious, "Satan Knew Eve" on a telephone booth. It was quite ominous and diabolical in its effect upon nighttime wanderers as he asked a passerby to legitimize what he saw and it was ascertained.

Gary felt growths like horns appearing on his head, though imagined, and tried to implant fear in the drivers who took notice of him. Our grand and powerful fellow thought he could control others with psychokinetic capabilities and hoped his brain would shrink through its convolutions but somehow maintain its profound strength to superintend the environment.

After releasing his incredible abilities, Gary felt an urge to look up from his lowered sights and noted a billboard highlighting his location. It read "New L heading Northwest." Gary thought he was being watched by others because he was strikingly attractive and it was well known that he had cosmic powers.

It was late at night and he walked nonchalantly into Plummer Park. He lay down and looking at the sky, noticed the Summer Triangle. Staring at the heavens and using his puissant skills, he moved what he thought were the stars and pondered about having an astral body. Believing his dominion was becoming more pervasive, Gary Woode felt regaled with these exploits and felt less timorous; however he knew he would survive and felt relieved with that reflection. He slumbered without dreaming and woke-up the following morning shivering from the damp cold.

He stopped at a cafe for coffee. "I hear you," a pretty blonde waitress said and gave him a potful. Our fellow drank it black and overcame the numbness in his extremities.

Getting through the night in good standing, he decided to walk up Hollywood Boulevard to the observatory and meanwhile encountered a flock of pigeons who saluted him honorary mayor and he decided to talk with the townspeople. Unaware of his psychosis, Gary went from one group to the next shaking hands after which he was chased by a dog who retreated when he lay down in a brook.

Summer was almost over and it began to get dark earlier. Still damp from the plunge, he walked down off the hill and stopped at a supermarket hearing someone say "Soak up some Chinese wine."

Gary thought he had developed an acidic pH and hoped by buying some club soda he could neutralize it. He only managed to get cold drinking that cold soda.

Around 8:30 our protagonist walked through an unusual part of town as the buildings and streets became flowing colors. He looked at cars and was ignored but he believed he attracted notice nonetheless.

Gary was a reserved individual who had signed up with the draft, but was looking forward to studying at the university. He spent his summers at the beach, turning bronze and watching girls. But now he felt omnipotent by ingesting different substances and walking through town under the influence of them. Gary was headed for a fall and was precipitating it.

Our adventurer traveled cross-town and down Pice Blvd. Stopping in a central part of an intersection and observing a car compelled to avoid a collision by withdrawing and then proceeding around him, he felt all powerful and indestructible.

Roaming the extent of Pico, he stopped at a market for a pack of cigarettes. "What do you think of the draft?" he queried an employee. A disconcerted cashier fumbled for an intelligent response: "We're being infiltrated by Iranians 'cause you'd be crazy to live there."

Gary tried to comprehend the meaning of this remark, but couldn't associate it with his peculiar dilemma. He dismissed it as inappropriate. He left the store promptly and continued his peregrination, walking up Robertson. Stopping to look in a mirror at one of the many shops, our crazed hellion was shocked to see his visage but with blue eyes instead of his familiar dark brown. Astounded with that discovery, he crossed the street and heard a siren as a police car was chasing after some bandit. Gary thought about detonating the getaway car and instantaneously it exploded in an immense wall of fire.

All he remembered was hearing sirens in the background as he was roused by a nurse who was trying to administer an injection.

Our patient, now alert, looked down upon the restraints and observed an entrancing paradox. On the adjoining strap it read "For the Sane" in bold letters.

This description of mental illness is your typical state hospital stay.

THE VISIT

He was admitted two minutes past midnight, April the first. After being screened by the psychiatrist on call he was escorted to the locked security ward and to the dorm by the third-shift aide. A quiet night, our new admittee covered himself with a blanket and dozed off.

He was a comely man, tall and thin, with premature gray in his hair. At 23 years he was just divorced and had his own business detailing cars.

Morning materialized, the lights in the dorm buzzed on, and the sudden effulgence mildly jolted the patients to the conscious moment. The patrons, resisting their leaden torpor, proceeded to dress themselves as the attendant trooped through the two-roomed dormitory. At 6:15 the population was guided down the long hall to the dayroom. They were allotted two cigarettes before they opened the showers. Some patients chose to recline on the couches until breakfast, served at seven. Usually breakfast was comprised of individual boxes of cereal, two pieces of toast, juice, milk, and a variety of eggs with bacon or sausage given twice weekly.

After the morning meal the trays were bussed, and the patients loitered around the short hallway waiting for the attendants to reopen the large dayroom. The television was turned on, the pool and foozball equipment were brought in, and after a few acute-treatment patients played games and socialized, the medication cart was wheeled in. Meds were administered by the day-shift nurse, and considered part of each patients individual treatment plan. Every patient on the ward was on medication. The nurse, a friendly fellow, initiated dialogue with our new patient.

"Welcome to our little world. Hi, I'm Mike, the day-shift nurse."

The nurse was a powerful-looking man with a red beard and hair and an engaging disposition that made others feel comfortable.

"The psychiatrist prescribed some medication. You're on your way to better mental health."

Our new patient looked up but couldn't formulate a response and instead started to tremble. The nurse realized this and handed him the Haldol spanule.

After the meds were dispensed there was about an hour until the doctors, psychologists, social workers, and therapists had their morning meeting. None of the patients knew what was discussed during the meetings, but it was certain these professionals wanted to help.

When 9:00 showed on the day-room clock, those patients with grounds privileges were allowed off the ward. The rehabilitation building contained a coffee shop, recreational facilities, and various treatment programs. Some patients were engaged in sheltered-work programs, some in the morning, some in the afternoon, and the more industrious out-patients worked both shifts.

One treatment program attracted the majority of the patients. This treatment mall was characterized by a friendly card-game, the latest magazines, for reading, television, radio and decaf coffee. The mall was a benign and insouciant ambiance and attendance there weighed favorably in each patient's file. Patients were expected back on the ward at 11:30 A.M. Meanwhile on the ward, our new client was approached by the ward psychologist. Her name was Sheila.

"If you'd like to talk, let me know," she said solicitously.

The new patient, whose name was James, nodded compliantly, but didn't utter a word.

At 12:00 P.M. lunch was served. Today's lunch was spaghetti with meat sauce. Desert was fruit cup. Each patient had one serving although some thinner patients had a double order.

After the midday meal, the dorms were opened and practically the whole ward lay down until 1:00 P.M.

At 12:30 P.M. the psychologist walked through the sparsely populated ward looking for group therapy prospects. When approached by this attractive woman, James consented.

"You can wait by the rec room," Sheila offered. "We're going to round up some of the other patients."

The group sat down around a large table. Sheila initiated the discussion by mentioning that what is discussed in the meeting is not to leave the meeting. Presently the patients introduced themselves. James introduced himself in a timorous manner.

"Feel free to talk about what may be troubling you or what comes to mind," Sheila addressed the gathering. She nodded approvingly toward James.

"I'm here because the psychiatrist feels I'm paranoid. But I know the real truth. Someone wants to kill me," James proffered brazenly.

"I don't know that. What makes you say that?" she queried gently.

"It's something I know. There is no way to prove it otherwise."

"That sounds logical, but what are you basing your knowledge of that on?"

"I'm basing my knowledge on a factual event."

Group therapy ended with most of the patients discussing their usual motley of topics. It was later that psyche interns interviewed some of the newer patients as a matter of procedure.

There were three students from the university: one male who wrote a check for James and two female residents who provided the clinical examination. The interns questioned James on a number of psychological assessments and during the interview James looked up and said extemporaneously, "I found out where the desert moon is."

"Where?" intervened one of the interviewers.

"Somewhere between here and eternity," was the answer. This intern seemed impressed by the rejoinder.

The examination ended, the check was signed over to the waiting patient, and the interns expressed their gratitude for this probe, all in the name of science. James thanked them for the honorarium and felt a numb gratification for the notice he received.

The following day James was interviewed by the team who were reviewing his treatment plan.

"You know," commented one of the staff, "We can't grant you privileges in this state."

"What state? New York or Kentucky?" James snapped.

"We're trying to help you understand the seriousness with which your beliefs are not wholly grounded in reality."

"How could they be grounded when I'm on a locked ward four floors high in a brick building? If I was outside then they might be grounded."

"We'll discuss his further. You can leave the meeting, James," the doctor motioned. The meeting moved on. They continued with their usual methodology in diagnosing and assessing the patient.

"He seems in touch with reality. His ideations are bizarre but he doesn't seem to exhibit feelings of persecution."

"His reasoning is logical and connected. He's not psychotic nor depressed yet he's preoccupied with paranoid beliefs. His diagnosis should be acute paranoia. Treatment would comprise psychotic therapy, group therapy, and rehabilitation programs."

The available staff would proceed to monitor and assess his progress by a number of indications, some of which are bodily motions, mood, and social encounters with other patients.

About a week later, the patient still exhibited signs of isolation, flightiness of ideas when approached verbally, and some rigidity, although that might have been attributed to the medication.

The team upon meeting again to address the patient's treatment plans and progress decided to confer privileges to James but with staff supervision.

Spring was unfolding on a temperate day, sunny and breezy; some patients went for a walk guided by two of the ward employees.

The surroundings to this institution were placid, sparsely sylvan, and the buildings outmoded. The patients, more notably those who were off the ward for the first time, seemed enchanted and somewhat elated with this modicum of freedom.

The group ambled down a slight grade to a set of swings and some patients momentarily regressed to a younger age and enjoyed themselves swinging in an extended arc and testing their fearlessness.

After a thirty minute outing the group returned to the ward experiencing a minor descent from their stimulating jaunt.

At 3:15 P.M. it was between shifts and all employees were meeting for transfer until 3:30 P.M.

A yell was heard down the long hall and immediately staff were seen to sprint toward the end of the hall. "Call a code" was heard above the commotion. Presently a team of medics with life-support equipment arrived. The patient, who was suspended by his belt to a large pipe, was lowered and administered to by the medics.

Some psychiatrists were summoned to the ward and filled out an incident form and forwarded it to a local hospital where James was given intensive care.

References

Davison, C., & Neale, M. (1996). Abnormal Psychology. Revised Sixth edition. New York: John Wiley and Sons, Inc.

Dyer, W. W. (1976). Your Erroneous Zones. New York: Funk & Wagnalls.

Garrison, K.C. (1948). The Psychology of Adolescence. Third Edition. New York : Prentice-Hall, Inc.

Jastrow, J. (1946). Freud-His Dream and Sex Theories. Cleveland: The World Publishing Company.

The Life Extension Foundation (1998). Disease Prevention and Treatment Protocols, Second Edition. U.S: William Faloon.

Slater, Philip (1976). The Pursuit of Loneliness. Revised Edition. Boston: Beacon Press

Wade, C., & Tavris, C. (1993). Psychology. Third Edition. New York: Harper Collins College Publishers.

Webster's New World Dictionary (1997 edition).

Webster's Ninth New Collegiate Dictionary (1992 edition).

www.ingramcontent.com/pod-product-compliance
Lightning Source LLC
Chambersburg PA
CBHW030812180526

45163CB00003B/1256